Henry Fanshawe Tozer

Researches in the Highlands of Turkey

including visits to Mounts Ida, Athos, Olympus, and Pelion, to the Mirdite Albanians and other remote tribes

Henry Fanshawe Tozer

Researches in the Highlands of Turkey
including visits to Mounts Ida, Athos, Olympus, and Pelion, to the Mirdite Albanians and other remote tribes

ISBN/EAN: 9783742875341

Manufactured in Europe, USA, Canada, Australia, Japa

Cover: Foto ©ninafisch / pixelio.de

Manufactured and distributed by brebook publishing software (www.brebook.com)

Henry Fanshawe Tozer

Researches in the Highlands of Turkey

CONTENTS OF VOL. II.

CHAPTER XVIII.

OLYMPUS.

Fear of Cholera — The Lazzaretto — Peculiar Theory of the Epidemic — Departure for Olympus — Katrin — Magnificent appearance of the Mountain — Ruins of Dium — Scala of St. Dionysius — Ascent to the Monastery — Description of it — Sawmills of Olympus — Disappearance of Bears — Dionysus and St. Dionysius — Ascent of the Mountain — The Central Basin — Chamois — Peak of St. Elias — View from it — Mythology of the Mountain — The Highest Summits — Snowstorm — Difficult Descent .. .. Page 1

CHAPTER XIX.

THE LOWER OLYMPUS.

Letochoro — District of Pieria — Leftocarya — Monastery of Kanalia — Monastic Contentment — Western Side of Olympus — Monastery of Sparmos — Suspicions of the Monks — Lake of Nezero — Mixture of Races on Olympus — View of the Plain of Larissa — Battle of the Gods and Titans — Clefts and Armatoles — Their Origin and History — Encroachments of the Turks — Clefts of Olympus — Popularity of the Clefts — Vigorous Measures of Ali Pasha — National Movement — Disbanding of the Armatoles 29

CHAPTER XX.

TEMPE AND OSSA.

Descent to the Peneius — History of the Koniarates — Site of Gonnus — Ambelakia — Its former Commercial Prosperity — Causes of its Downfal — Vale of Tempe — "The Beauty's Tower" — Ancient Descriptions of the Pass — Ballad of the Salamvria — Monastery of St. Demetrius — Unfavourable Reception — Handsome Central Church — Commencement of Decay — Seacoast of Ossa — Plain and Town of Aghia .. 59

CHAPTER XXI.

THE VRYKOLAKA, OR EASTERN VAMPIRE.

Apparitions of Spirits — Terror inspired by the Superstition — Origin of the name Vrykolaka — Connection with Lycanthropy — Malignant Vampires — Frequently produced by Excommunication — The Body undecomposed — Effect of Absolution — Stories relating to it — Modes of Exorcism — Supposed Offspring of Vrykolakas — Burning the Body — Tournefort's Narrative — Innoxious Vampires — Ballad of "The Spectre" .. .. Page 80

CHAPTER XXII.

PELION.

Meeting-point of Ossa and Pelion — Ruins at Skiti — Chain of Fortresses on Pelion — Ferocious Dogs — Ali Pasha regretted — Woods of Polydendron — Ruins of Casthanæa — Keramidi — Ascent to the Ridge — Places named from Trees — Plain and Mountains of Thessaly — Intelligent Greek farmer — Views of the Eastern Coast — Zagora — Hospitable Reception — Prosperity of the Townships of Magnesia — Romaic and Neo-Hellenic Languages — Dancing the Romaika — Inferior Position of Women 98

CHAPTER XXIII.

PELION *(continued).*

Superb View of Athos — Ancient and Modern Vegetation of Pelion — Ascent to the Summit — View from it — Cave of Chiron the Centaur — Descent to Portaria — Site of Iolcos — Hill of Goritza — Ruins of Demetrias — Volo — Ill-treatment of Bulgarian Labourers — Excursion to Milies — No Turks on Pelion — Homer and Virgil on "The Giants' Mountains" — Olives and Oil — Library and School of Anthimos Gazes — Quarantine 121

CHAPTER XXIV.

THESSALY AND METEORA.

Journey in 1853 — Departure from Salonica — Approach to Thessaly — Larissa — A Grandee of the Old School — Plains of Larissa and Tricala — Tricala — Modern Adaptations of Ancient Names — Rocks of Meteora — The Great Monastery — Admittance Refused — Monastery of St. Stephen — Remarkable View over the Plain of Thessaly — Fear of Robbers — Monastery of Barlaam — Ascent by the Rope and Net — Ballad on the Abbot of Barlaam .. .. Page 140

CHAPTER XXV.

PINDUS.

Upper Valley of the Peneius — Malacassi — Defeat of the Greek Insurgents in 1854 — Pass of Zygos — Fountain-head of Rivers and Meeting-point of Races — Metzovo — The Wallachs — Origin of the Name — Evidences of their Roman Descent — Customs and Beliefs — Roman Colonies in Dacia — Migration south of the Danube — Subsequent History — Bulgaro-Wallachian Kingdom — The Wallachs of Thessaly — Present Condition 163

CHAPTER XXVI.

YANINA AND ZITZA.

Valley of the Arta — Races of Southern Albania — Yanina and its Lake — Oracle and Sanctuary of Dodona — Excursion to the Monastery of Zitza — Byron's Description of it — Albanian Rifle Practice — Scene of Ali Pasha's Death — His Serai and Tomb — Notice of his Character and Career — Atrocious Massacre of Greek Women 183

CHAPTER XXVII.

SULI.

Frequency of Thunderstorms in Epirus — Greek Theatre at Dramisius — Ruins of an Hellenic City — Mount Olytzika — The Molossian Dog — Sources of the Charadrus — Mountains of Suli — Moonlight Views — Kako Suli — Ruined Homes of the Suliotes — Their History — Invasion of their Country by Ali — Samuel the Caloyer — Their Expulsion — Gorge of the Acheron — The Palus Acherusia — The Inferno of the Greeks — Parga — Its Cession to the Turks — Ballad on the Subject 199

THE HIGHLANDS OF TURKEY.

CHAPTER XVIII.

OLYMPUS.

Fear of Cholera — The Lazzaretto — Peculiar Theory of the Epidemic — Departure for Olympus — Katrin — Magnificent appearance of the Mountain — Ruins of Dium — Scala of St. Dionysius — Ascent to the Monastery — Description of it — Sawmills of Olympus — Disappearance of Bears — Dionysus and St. Dionysius — Ascent of the Mountain — The Central Basin — Chamois — Peak of St. Elias — View from it — Mythology of the Mountain — The Highest Summits — Snowstorm — Difficult Descent.

WE found Salonica in great excitement. The cholera, — which during the early part of this summer had seized, one after another, on all the great ports of Turkey, beginning from Alexandria, and was now raging frightfully at Smyrna, Constantinople, and the Dardanelles, — had made its appearance in the *lazzaretto*, which was consequently full of dying persons. As yet it had not attacked the town, and the quarantine station was at some distance off at the head of the harbour; but the inhabitants were naturally afraid that it would soon be amongst them, unless the strictest prohibitive regulations were observed. The day after our arrival a great scene occurred in consequence of the arrival of a steamer from Constantinople with several dead on board, and more infected. As there was no room in the *lazzaretto*, it was greatly feared that they might be admitted into the place; and when the Pasha came down to the harbour to investigate the matter, he was surrounded by hundreds of the Jews, of whom so large a part of the population is composed, who stood round about wailing dismally, and beseeching that the vessel might be sent away. This was ultimately done; orders were given that she must leave the port, it being added that she might go where she liked, as long as she landed no one here.

The story of the building of this *lazzaretto* added another to the numerous instances of Turkish mal-administration which were continually coming to our ears; it is wearisome to be so frequently mentioning them, but without doing so I should not be giving a faithful picture of the condition of the country. The old *lazzaretto* was a narrow unhealthy building situated near the town. The last time the cholera visited the place, when numbers of persons were dying in this den, the Consuls made a representation to the Pasha, urging that a larger and better station should be built at some distance off. The Pasha replied that he must communicate with the authorities at Constantinople, and telegraphed accordingly; on which the answer was returned that nothing should be done. Knowing that in this country everything must be accomplished by importunity, the Consuls again sent in a memorial, which was followed by the same course of proceeding as before, concluding with another negative on the part of the Government. At last the Consuls tried another tack, and offered this time to erect the building at their own expense: after this the authorities made no further objection, and the present station was built at a cost of 120*l*.

During our stay in the city on this occasion a theory on the subject of cholera was propounded to me, which may possibly be new to medical practitioners. I had gone to the Austrian embassy to enquire for letters; for in Salonica, as at Constantinople, there is no central postal administration, but it is necessary to apply at the office of the line of steamers, or the overland route, by which your letters may be expected to come. While waiting there I got into conversation with the porter of the embassy, an old Greek, and of course among other topics we discussed the unfailing one of the approaching malady. He told me that during thirty years or more that he had resided in the place he had seen many very severe visitations of cholera, and that he had constantly observed that the worst parts of the town were not those that suffered most from it. "Do you mean to say," I rejoined, "that it seems to attack by preference the dwellings of well-to-do people?" "Exactly so," he said; "the well-built portions of the town, where there are airy houses, extensive gardens, and some attempt at drainage, are the most exposed to it, while the filthy crowded quarters of the lower-classes are more commonly spared." "Well," I remarked, "what you mention is very curious, and contrary to our usual experience; have you anything to suggest by way of explaining it?" "Well, sir," he replied, "I don't pretend to set up my view against anybody else's, because I may be wrong; but my idea is that in the lower part of the town the atmosphere is so foul that the cholera can't get in." On hearing this singularly *naïve* explanation, I could not help recalling the Italian proverb, "l'un diavolo caccia l' altro."

We had hoped on our arrival to spend two or three days at Salonica; however, as there was some cause of fear lest on account of cholera the shores of Olympus should be put in quarantine against the city, it seemed better to be on the move again after one day's interval, which was spent in business and in visiting our friend Mr. Wilkinson. Accordingly, on the 6th of August we started once more, having chartered a vessel to convey us down the bay, a six-oared Smyrna caique, quite elegant in her appointments as compared with the ordinary lumbering market boats and coasters of these seas, and a tight little craft withal, for though not more than six feet in width and without a deck, she had made a voyage to the Crimea during the war. As we crossed the bay, we passed, some way from the land, through the stream of the Vardar, whose pale and turbid water is distinguished by a clear line of demarcation from the deep blue sea into which it runs, a peculiarity which may also be noticed at the point where the Rhone enters the Mediterranean: as we emerged from it on the opposite side the contrast of colour was less strongly marked, probably in consequence of the quantity of fresh water which is contributed to that angle of the bay by the Vistritza (Haliacmon). The breeze was in our favour; and the splendid forms of Olympus and Ossa rapidly enlarged before us, until a little before sunset we arrived at a *scala* or landing-place to the eastward of the village of Katrin, which lies in the middle of the plain on the northern side of Olympus. An Egyptian war-steamer was lying off the coast; and when we enquired the object of her visit, we found that she was taking in coal, which was brought from a mine lately opened at a place called Triandista, at the foot of the mountain, though apparently the operations had not as yet been carried much below the surface. At the landing-place were lying large piles of sawn timber ready for exportation; and on the beach a market boat was half stranded, laden with water-melons intended for the market of Salonica, but now forbidden fruit, and a dead loss to the owners, as the authorities of that place had interdicted every kind of raw vegetable. When we had obtained mules, we rode through the moonlight over the plain to Katrin, where we were hospitably received by the brother of the dragoman of our consulate at Salonica, who is an important man here, an Epirote from Yanina,[1] Nicola Bizko by name. In his possession we noticed two objects, which were of some interest to us at the time; the one a pair of chamois' (ἀγριογίδια) horns from Olympus, where he said those animals abound; the other a cigarette holder made by the shepherds of that mountain, who are skilled in the art of wood-carving. M. Bizko expressed his warm admiration of Mr. Wilkinson, which seemed chiefly to be excited by the freedom and candour with which that gentleman criticized the character of the Turkish grandees; for he repeated the following sentence over and over again with great gusto: "If a Pasha is a good fellow, he calls him a good fellow; and if he isn't, why he calls him an ass (γαΐδερο)."

The next morning we started for Olympus. To the westward at this point a range of mountains bounds the view at no great distance, forming the extreme end of the Bermian chain, the most striking point in which is

[1] The name of another Epirote, in addition to those already mentioned, leads me to speak of the reason why so many persons of this race are found at a distance from their homes. In the 'Times' for April 11, 1868, under the heading "Life in Epirus," there is a notice of a pamphlet lately published at Athens, and entitled 'The East and the West,' in which it is stated, and the statement confirmed by statistics, that the reason why so many Christians migrate from Epirus and trade in other parts of Turkey is, that the excessive burden of taxes in that province forces members of families to leave their homes, in order, by their gettings elsewhere, to maintain those who remain behind.

the summit that was called Mons Pierus in ancient times;[2] but everything is dwarfed by the great mountain of the Gods, which lies directly to the south, and presents an indescribably grand appearance, as it rises at once its whole height of 10,000 feet immediately from the plain, with steep precipices in its upper parts, and below innumerable buttresses, divided again and again into minor ridges and valleys, thickly clothed with feathery woods. It formed a striking contrast to the broad and not delicately marked forms of the mountains we had hitherto seen on this journey, especially to those of the Scardus; and exactly realizes what is expressed by the Homeric epithet, "many folded" (πολύπτυχος). The pyramidal summit which from this side appears the highest, though it is not so in reality, is that of St. Elias. We rode along the plain until we passed a small river, running in a deep bed, to which the natives gave the name of Kryonero, or Coldwater; shortly after which, near the village of Spighi, we noticed on a low hill a large tumulus, with trees growing on it; this, according to Leake's suggestion, may perhaps represent the site of the city of Pieria, the principal town of the district of the same name, which comprised all the country that lay under the northern and eastern slopes of Olympus. Our first halting-place was the *kalyvia*, or huts of Melathria, in the neighbourhood of which are a number of plentiful clear springs, with picturesque environs from the irregularity of the ground where the slopes first begin to rise towards the mountain, and the numerous trees, principally oaks, which are festooned with clematis and wild vine. Close to this spot are the ruins of the city of Dium, which are scattered over a wide space of ground, though

[2] Plin., iv. 15.

but little is traceable. Here and there may be seen a drum of a white marble column and a few other squared blocks, with lines of raised mounds and occasional walls; in one place the shape of the ground, forming a wide trough with semicircular ending, seems to indicate the position of a stadium. The theatre, though overgrown with trees, is well marked, from the high bank which was built to support the seats; it faces north-east, and is of considerable size, for the *scena* is not less than 150 feet across. Situated as the city must have been on a gentle slope close to marshy ground, it presents us with the same peculiarity which is so noticeable in Pella—a place of great importance, built on a site which was neither strong nor healthy.

From this place the ground begins to rise, in the direction of Olympus, with a gradual incline through prickly holly-oak and thorny palluria, vegetation not favourable to the traveller's clothes. At a point where this moderate ascent meets the steep slopes of the eastern side of the mountain, is the *scala*, as it is called, of the monastery of St. Dionysius, a farm which serves as an emporium, lying halfway between it and the sea; while the *scala* of St. Theodorus below (more properly so called, for a *scala* generally signifies a landing-place), serves as a port both for the monastery and for the neighbouring village of Letochoro, which is the principal place of these parts, and the residence of the Derven Aga, or guardian of the mountain passes. The upper *scala*, at which we now arrived, is a building of solid construction, and from what we afterwards saw appears more inhabited and better provided with the means of subsistence than the monastery itself. Here we found the hegumen, to whom we had a letter of introduction from Bizko, a good-natured man, though on Athos we should have called him a very common-place monk ; he regaled us with a dish of mixed eggs and cheese, and provided us with some of his own mules to continue our journey, suggesting that they would be safer among the rocks than the horses which we had brought with us from the plain. Mounted on these strong and well-fed animals, we ascended the extremely steep mountain side behind the *scala*, by rough zigzags, meeting on our way strings of other mules laden with planks, which had been sawn in the upper valleys. Down this abrupt slope water is conveyed to the *scala* by means of a covered watercourse; for except in the valleys there are few springs on the sides of Olympus, and the building stands at the foot of a ridge, up which the road to the monastery has to be carried, as the sides of the neighbouring ravines are too steep to admit of its passing along them.

Throughout the whole of this ascent the vegetation is of the most luxuriant description ; in the lower part it is mainly composed of holly-oak, catalpa, and two kinds of arbutus, the less common of which is easily distinguishable in spring-time by the bright red colour of its stems ; further up the place of these is taken by oak, beech, and several kinds of light green pines and firs. The views were such as are only found in the neighbourhood of the Ægean, forming a great contrast to all that we had lately seen, and forcibly reminding us of Athos and other places we had visited in former years, from the tender blue of the sea, the southern vegetation, and the sharply cut outline of the mountains. Towards the sea lay the dry slopes of Pieria, in the upper part of which was seen Letochoro, a place of some size, distinguished by its red-tiled roofs, on the side of a deeply embedded torrent ; further south appeared the white walled castle of Platamona, with its conspicuous tower crowning a low height; then the mouth of the Peneius, with alluvial land stretching out into the sea; and beyond, the shoreless cliffs of Ossa and its glorious gray peak. To the north was the level country about Katrin, diversified in parts by extensive tracts of woodland; and on the opposite side of the water lay the three promontories of Chalcidice, one behind another, the peak of Athos rising supreme above all; and beyond the extremity of these was the distant island of Aghiostrati to the south of Lemnos. In this part of the ascent the rocks are composed of a very white limestone, and when we reached the ridge we found it extremely sharp, so that it resembled the narrow knife-edges which are so characteristic of the Pyrenees. On the north side three similar ridges were seen, with deep valleys between them, covered with trees and excessively beautiful; while to the south appeared a wilder gorge, along the side of which we rode at a great height above the stream, with views continually opening out of the higher peaks, which were grey above and dark with firs below. At last the great basin disclosed itself which forms the heart of the mountain, closed in on the two sides by lines of magnificent summits which are joined by a saddle at its western end. At the entrance of this lies the monastery. As we began to descend towards it night came on, but the full moon rose in splendour from the Ægean, and lighted us for the remaining hour of our journey. In one part of our route we had met five-and-twenty Palikars (guards) returning from an expedition in search of a band of Clefts, who were said to be abroad on Olympus, and shortly after our arrival two of them made their appearance at the monastery, having been sent by the Derven Aga for our protection; these we kept, partly because we could not well help ourselves, and partly with the idea of making them useful as guides or porters in the ascent of the mountain.

The monastery of St. Dionysius, one of a number of very ancient monasteries on the sides of Olympus, is situated in the midst of walnuts and other trees, about 100 feet above the stream which flows eastwards from the heart of the mountain through the deep valley, along the sides of which we had approached. It is 3080 feet above the sea. The surrounding scenery is superb, and could not easily be surpassed in Switzerland, from the grandeur of the rocky basin in which it lies, whose steep sides are overlooked by the loftiest summits, rising to the height of 7000 feet above the valley. In contrast to this sublime desolation, the lower slopes present the refreshing sight of varied and delicate green foliage, with clear water flowing among the hard white rocks. The first court of the building, as you enter, is in ruins, having never been restored since it was destroyed by the Turks in 1828; the inner court forms an irregular square, with corridors running round it supported on two rows of arches, while the greater part of its area is occupied by the church, a building with five cupolas and a lead-covered roof, but without any appearance of antiquity, and containing nothing of interest inside, except a picture of St. Dionysius of some merit. The guest-chamber, which we occupied, was a nice clean room overlooking the valley. The whole number of monks who belong to the monastery is eight or nine, and at their head is a single Hegumen, though they do not on that account call themselves a Cœnobia, as would be the case on Athos. During the summer most of these are generally absent on business, being engaged in looking after their property. At the time of our visit there was but one monk there, and he only a *candelaptes*, or lamplighter; and as there was no priest, there could be no service. Before our arrival we had heard great things of the good cheer that would await us; and one of the attendants who had accompanied us from the *scala*, when we questioned him on the subject, replied by putting his hand to his throat, as if measuring the contents of a bottle, and saying "full up to here!" Our expectations on this point, however, were doomed to disappointment, and it was as much as we could do to obtain sufficient to subsist upon during our stay. The revenues of the monastery are in some small measure drawn from cultivated land in the neighbourhood, and from property in Russia, but by far the greater part of them are derived from the sawmills (πριόνια), of which the monks possess a great number on the mountain sides. These, together with the right of felling timber in the forests, they let to woodcutters, who come every year from Albania for the purpose; not, however, from the Dibra, like those I have already mentioned, but from the neighbourhood of Castoria. Within each mill there is a circular saw, worked by water, and extremely rude; but in this country, where anything like machinery is so rare, they are widely known as triumphs of art. From them and the timber an annual revenue of 40,000 piastres (about 330*l*.) is obtained; and the wood is exported to Smyrna, to Constantinople, and even to Alexandria.

On waking the next morning we heard Jove thundering above our heads,—a propitious sound for us, since we knew, from former experience on Eastern mountains, that except after rain there would be no chance of a clear view at this time of year. At Salonica we had been told that it was expected soon, as none had fallen for six weeks; and now the thunder brought it, and it continued to fall with short intervals during the whole day. We were not sorry to be obliged to rest awhile after so long continuous travelling, and spent the day in wandering about the valley in full enjoyment of the unwonted luxury of cool air and fresh breezes, and in making enquiries about the mountain and our prospects of ascending it. In the evening we were able to appreciate a blazing fire on our hearth. In the course of our enquiries we came upon a very curious legend, which was related to us by the old monk. Having asked about the wild animals which were met with in the forests, we were informed that there were deer, wolves, and various other kinds, but no bears ; to account for the absence of which the following story was related. Once upon a time, when St. Dionysius, the founder of the monastery, was ploughing on the mountain, he was called off from his work, and forced to leave his ox with the plough in the field. During his absence a bear came down and devoured the ox ; but the saint on his return, discovering what had happened, seized and harnessed the offending beast, and made him drag the plough. After which time the bears (considering apparently that such treatment was *un peu trop fort*) disappeared from Mount Olympus.[3]

---

[3] In M. Heuzey's book, 'Le Mont Olympe et l'Acarnanie,'—which, together with Dr. Barth's travels, already referred to, are our principal authorities on this mountain,—there is a slightly different account of the legend, the bear being represented as devouring the saint's horse, and being condemned to carry the saint instead ; the result in both cases is the same. M. Heuzey's and Dr. Barth's books are as complete a contrast to one another as the narratives of a Frenchman and a German usually are: the former is thoroughly inaccurate, the latter minutely faithful. As Dr. Barth's volume is the later in date, he had the opportunity of administering such castigation as a German never spares to a Frenchman, in doing which he has, perhaps, too much ignored M. Heuzey's carefulness in investigating the inscriptions and antiquities, such as they are, of this neighbourhood. But the map that accompanies the Frenchman's volume deserves

The reader's mind will naturally revert to St. Patrick's expulsion of the snakes from Ireland; and in Crete, the same office of freeing the country from noxious animals is attributed to St. Paul, as in ancient times it was to Hercules.[4]

Though the monastery bears the name of St. Dionysius, yet in reality it is dedicated, like all the monasteries on Olympus, to the Holy Trinity; and it is not improbable that the reason of this may have been to transfer the associations of the locality from the supreme heathen deity to the Christian Godhead. The founder seems to have come originally from one of the monasteries of Meteora in Thessaly, and to have established this convent in the twelfth century; but it is possible that an older building dedicated, like so many other edifices of the Greek church, to St. Dionysius the Areopagite, may have occupied the same position.[5] Dr. Barth suggests the idea that an altar of Dionysus may have stood here in heathen times;[6] and the white marble slabs which are found in the building, together with the mouth of a well made out of a piece of a column, seem to suggest that it was a classical site; to which it may be added that the worship of Dionysus is known to have existed on Olympus.[7] And that the ideas connected with the ancient god and the Christian saint are confused in the minds of the people is shown by the following remark-

---

the severest animadversion, for the northern half of the mountain, including all the highest summits, is as falsely delineated as if the compiler had never been on the spot. Unfortunately, as the work was published by the French Government, these mistakes have found their way into other maps.

[4] Pashley's 'Travels in Crete,' ii. 261.
[5] Heuzey, p. 130.   [6] Barth, 'Reise,' p. 199.
[7] " Μάκαρ ὦ Πιερία
σέβεταί σ' Εὔιος."—Eur. 'Bacch.,' 565. See also infra, p. 31.

able story, which was taken down by a German Professor[a] from the lips of a Bœotian peasant :—

"Dionysius, when yet a child, was travelling through Greece on his way to Naxia; but, as the journey was very long, he became tired and sat himself down upon a stone to rest. Now while he sat thus with his eyes fixed on the ground, he saw a tiny plant spring out of the earth at his feet, which seemed to him so beautiful that he at once determined to take it with him and plant it. So he pulled it up and carried it away with him; but, as the sun shone with great power, he was afraid it would wither before he reached Naxia. On this he found the bone of a bird, into which he put the little plant, and continued his journey. But in his holy hand it grew so fast that it soon projected from the bone both above and below. Again he was afraid of its withering, and bethought himself of some further device. So he found the bone of a lion, which was larger than the bird's, and into this he put the bird's bone which contained the little plant. Before long, however, it grew out of the lion's bone also. Thereupon he found the bone of an ass, which was still larger than that of the lion; into this he put the little plant with the bird's and lion's bone, and so arrived at Naxia. But when he was going to plant it he found that its roots had twined themselves close round the bones of the bird, the lion, and the ass; and accordingly, as he could not pull it out without damaging the roots, he planted it just as it was, and at once the plant sprang up on high, and, to his delight, bore the loveliest grapes, from which he proceeded to prepare the first wine, and gave it to men to drink. Then what a marvel did he behold! When men drank of it, at first they sang like birds; when they drank more of it, they became strong as lions; and when they drank still more, they became like the ass."

On the following morning (August 9) the rain had ceased; and though the clouds still hung low in dense masses, yet as our time was precious, and there was a chance of their clearing off, we determined to start for the ascent. At a quarter past nine we were off, accompanied by a young Greek from the monastery, an active obliging fellow; and our two Palicars, one of whom was

---

[a] Professor Siegel, by whom it was communicated to Von Hahn, and published in his collection of 'Griechische und Albanesische Märchen,' vol. ii. pp. 76, 77.

a heavy unimpressive creature,—the other vain, officious, and talkative, always obtruding his advice when it was not asked. One virtue, however, this second one had; he was a great hand at roasting a kid's entrails round a ramrod, a dish which is deservedly esteemed a great delicacy. He had already shown us his proficiency in this art at dinner-time on the previous day, when he appeared with the first course on this quaint (but in this country approved and orthodox) spit, from which, after dividing it into portions with his yataghan, he projected it on to our respective plates. Both these guards were indifferent mountaineers, and delayed us much by frequently wanting to stop, complaining of their packs,—which, as they only contained some provisions and a few wraps in case of a night bivouac, weighed about a quarter of what a Swiss guide would carry with pleasure. But *men* in the East have a rooted objection to carrying anything but a gun; "if we weren't loaded like this," they would exclaim, "it would be a different thing; as it is, we are doing the work of mules" (ἂν μόνον δὲν ἤμασθον τόσο φορτωμένοι· τώρα εἴμεθα εἰς τὸν τόπον τῶν ζώων). At first we ascended in a northerly direction, zigzaging up the steep slopes at the back of the monastery, until we attained the ridge, which here runs westward: from this point there were pretty views towards the sea, under the thick curtain of clouds, the extremity of the distant promontory of Pallene appearing above the mouth of the deep gorge. We then proceeded along the side of this ridge among thick woods, where in many places the path had been well marked by the wood-cutters; at their last station we stopped for breakfast at half-past eleven in the midst of the clouds, as there are no springs higher up on the mountain Next followed a long and steep ascent, in the course

of which we left the beeches below us, and at last the
firs also, at the height of 7300 feet, though a few stragglers
may be seen higher still; underneath the firs the box
bushes form a fine undergrowth. Not long after the
bare mountain side was reached, we found ourselves on a
small level, where a Wallach shepherd was tending his
flock of goats; it was surprising that they could find
sufficient pasture to browse on, for the ground was composed, as it is throughout the northern and eastern
districts of Olympus, of a hard white limestone, which
easily breaks into pieces, allowing only tufts of grass and
a few flowers to grow. Our companions were continually
urging us to be on our guard against the sheep dogs;
and with good reason, for they are extremely fierce, and
more than one traveller, when alone on the mountains,
has escaped from them only by sitting down in the
midst of them, as Ulysses is described to have done in
the 'Odyssey,' in which case they will surround you
without attacking you. The shepherd was despatched
in quest of snow (there being no water), and soon
returned with a huge lump of it, several pounds' weight,
which he carried over his shoulder, stuck through with
a pole.

At this point the clouds opened, and showed us the
position we had reached. We were standing on the
ridge which forms the northern boundary of the great
central basin, above the monastery of St. Dionysius, into
which the descent was extremely steep and almost unbroken, while on the opposite side the rocks rose with
hardly less abruptness to a somewhat lower elevation,
surmounted by a conspicuous peak, some five miles
distant from us, to which our companions gave the name
of Golevo. At the western end the line of cliffs sweeps
round in a grand curve, and forms the saddle which we

had seen from the valley, joining the two parallel chains of summits; and at the south-west angle is a cluster of peaks, well compared by Heuzey to the cupolas of a

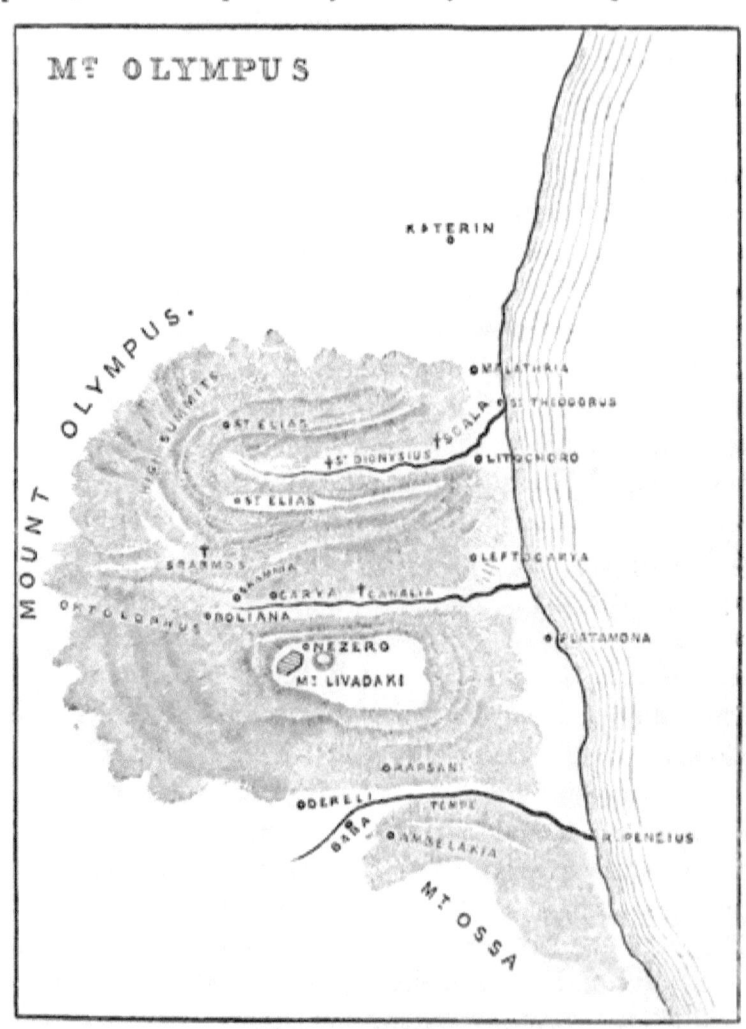

Mount Olympus.

Byzantine church, from which one stands out conspicuous, probably the St. Elias of the opposite range. Nothing could well surpass the magnificence of the enormous basin below us, filled as it was with masses of white cloud, swirling and seething as in a huge cauldron. This view at once showed us how inaccurately Olympus is delineated in all the maps, where it is made to run from north to south, which, indeed, it seems to do when seen from the sea; whereas in reality it is composed of two ranges which run from east to west, and meet at their western end. The ridge which we now followed is the same which we had seen cutting the sky in our view of the mountain from the plain of Katrin. A rapid ascent brought us to a narrow *arête*, from which the ground descended steeply on both sides; in crossing it our attendants, who had the mocassins of untanned hide which Greek mountaineers usually wear, preferred clambering along the loose débris under the rocks on the southern side, while we ourselves, being shod with strong boots intended for Alpine climbing, found the crest of the ridge more agreeable walking, notwithstanding the wind, which swept over it with great force from the north-west. After following this for some distance we reached a bare stony platform at the foot of the peak of St. Elias, in the midst of which lay a large basin filled with snow. Of this we saw considerable quantities in various places on the slopes of the peaks, and in the rifts of the mountain sides; but notwithstanding this, no part of Olympus is within the limit of perpetual snow. At the same time the Homeric epithet of "very snowy" (ἀγάννιφος) is thoroughly applicable to it, and I myself have seen it in the month of May presenting all the features of a snow mountain. "White as Olympus" is a comparison used in the modern Greek ballads;[9] and, in

---

[9] "ν' ἀσπρίσῃ σὰν τὸν Ὄλυμπο," in Passow, 'Popularia Carmina Graeciae recentioris,' No. 300, line 6.

numerous passages of the ancient poets,—such as that of Homer, where he calls it the "dazzling Olympus [10];" and that of Sophocles, where he speaks of its "gleaming radiance," [11] reference seems to be made to its snowy brightness. Indeed, it would seem to owe its name to this same attribute, if, as is most probable, Olympus signifies "the shining one." [12]

Before we commenced the ascent of this peak a herd of ten or twelve chamois appeared on one side of it; on seeing which our men made us sit down, and waited in hopes of getting a shot at them, though with their awkward Albanian guns they had small chance of sport. We had seen several of these animals before, when crossing the *arête*, and in another place we had put up a large covey of partridges; but though our Palicars thought they found traces of deer here and there, this was all the game we saw; indeed, though the mountaineers in Turkey

---

[10] "αἰγλήεντος Ὀλύμπου."—'Iliad,' i. 532.

[11] "Ὀλύμπου μαρμαρόεσσαν αἴγλαν."—Soph. 'Ant.', 610.

[12] Derived from λάμπω, according to Curtius, 'Grundzüge der Griechischen Etymologie,' i. p. 231. What makes this derivation the more probable, is the almost generic use of the name Olympus for a lofty mountain, or chain of mountains, in the same way as Ida (from ἴδη) is used for a wooded height. Thus the successive chains in the north of Asia Minor are called by that name; and it is more likely that they received it in the course of the western migrations of the Pelasgic races, than that the name was transplanted from the West, and transferred to them at a later period. It has often been remarked how the highest mountains in all parts of the world have received their names from their whiteness.

While on this subject I may refer to Colonel Leake's remark that "the name Elymbo, *i.e.*, Ἔλυμπος, which is now applied to the mountain, not only by its inhabitants, but throughout the adjacent parts of Macedonia and Thessaly, is probably not a modern corruption, but the ancient dialectic form, for the Æolic tribes of Greece often substituted the epsilon for the omicron, as in the instance of Ὀρχομενός, which the Bœotians called Ἐρχομενός." The fact here stated is true as regards the surrounding districts, and the philological explanation is not improbable; but on the mountain itself Ὄλυμπος, and not Ἔλυμπος, is the name used by persons of all classes.

often talk about game, yet the general absence of it throughout the country is very remarkable, as there is plenty of cover: during the whole of this journey we had seen nothing but a partridge or two and one hare, and when we came to talk to persons who had had experience of the shooting, they complained that they had had great fatigue and little sport. After this we mounted the steep peak, which is about 800 feet high from its base to the little chapel dedicated to the prophet Elias, which crowns its narrow summit; this is a rude structure of stones simply piled together, containing nothing inside, and without any sign of its being a sacred edifice. We arrived there at twenty minutes past four.

By this time the clouds, which had gradually been dispersing, had in great measure cleared away, leaving the distance remarkably distinct for the time of year: and though the entire panorama was never completely visible at one time, yet we saw the views clearly in almost every direction. To begin from the eastern side—far below us through the opening of the gorge of St. Dionysius, a narrow section of the Pierian lowlands was seen, and southwards of it, appearing over the seaward slope of the opposite ridge of Olympus, the peak of Ossa and its outline, reaching to the Ægean; directly to the south the distant view was barred by Mount Golevo and the other peaks which rise on that side, until a depression occurs to the south-west, through which a part of the plains of Thessaly appeared as in a vignette, overhung with clouds, and framed by the nearer mountains. Then followed other lofty summits close to us, beyond which, to the west, as far as the eye could reach, the numerous ranges of the Cambunian chain filled all the view; while at our feet was the entrance of the deep defile forming the pass of Petra, through which Xerxes entered Greece, with yawning chasms and impassable precipices descending towards it, the "barrier crags of precipitous Olympus" of the Orphic poet of the Argonautica.[13] To the northwest, within the line of the Bermian range, was seen far off a plain among the mountains, that which is drained by the stream of the Haliacmon; directly to the north other chains bounded the horizon in the direction of Vodena, within which the plains of Salonica and Katrin reached from the former city to the buttresses of Olympus. Salonica itself, at the head of the winding Thermaic Gulf, was clearly visible with the lofty Mount Khortiatzi rising behind it; then followed other distant mountains, and the lowlands of Chalcidice, with its three remarkable promontories, the isthmus of Pallene appearing almost like a thread, while Athos rose majestic above all. Far in the distance was seen the island of Agiostrati, and another further to the south, one of those which run off from Mount Pelion. A magnificent view indeed it was, together with the wide expanse of sea, which on this day was in colour a delicate soft blue, and the wonderful deep basin of the heart of the mountain in front, into whose inmost recesses, nearly 7000 feet below us, we could look down. The weak point in it was that the country on one side was almost wholly excluded from view, and this must necessarily be the case in all views from Olympus, in consequence of its double line of summits.

The heights on which we were standing were no unworthy position for the seat of the Gods. Here, according to the earliest Greek mythology, they dwelt in a serene ether, of which Homer sings that "it is never disturbed by winds, nor bedewed by showers, nor invaded by snow, but a cloudless atmosphere is there outspread, and a clear

---

[13] "Οὐλύμπου δὲ βαθυσκοπέλου πρηῶνας ἐρυμνούς."—line 462.

white brilliancy pervades it."[14] Below it lay the clouds, which separated it from the terrestrial region, and were regarded as the gates of heaven, being entrusted to the guardianship of the Hours, whose duty it was to roll back and close again the cloudy portal.[15] On the highest peak was the palace of Zeus, constructed for him by the skill of Hephæstus, where he was wont to summon to his council-chamber the other gods who dwelt below, on the flanks and in the recesses of the mountain.[16] At a later time this patriarchal conception of the dwelling-place of the deities passed away, and the name of Olympus was used for the most part in a metaphorical or ideal sense, to signify the heaven above. But notwithstanding this, and though it was situated at the extremity of Greece, Olympus never lost the pre-eminence and sacred character, as the home of the Gods, which it acquired at an early period, when the Hellenic race inhabited the neighbouring country, as seems to be implied by the fact that so many of the most primitive Greek legends, such as those of the battle of Gods and Titans, of Prometheus and Deucalion, and of the Argonauts, are connected with Thessaly. Its great elevation, 2000 feet higher than any other mountain of Greece, would of itself mark it out as an object of especial veneration, from the tendency which was common among the early races to attribute a

[14] " Οὐλυμπόνδ', ὅθι φασὶ θεῶν ἕδος ἀσφαλὲς αἰεὶ
ἔμμεναι· οὔτ' ἀνέμοισι τινάσσεται οὔτε ποτ' ὄμβρῳ
δεύεται, οὔτε χιὼν ἐπιπίλναται, ἀλλὰ μάλ' αἴθρη
πέπταται ἀνέφελος, λευκὴ δ' ἐπιδέδρομεν αἴγλη."—'Od.,' vi. 24 *sqq.*

[15] " αὐτόμαται δὲ πύλαι μύκον οὐρανοῦ, ἃς ἔχον Ὧραι,
τῇς ἐπιτέτραπται μέγας οὐρανὸς Οὔλυμπός τε
ἡμὲν ἀνακλῖναι πυκινὸν νέφος, ἠδ' ἐπιθεῖναι."—'Il.,' v. 749 *sqq.*

[16] "———— ἧχι ἑκάστῳ
δώματα καλὰ τέτυκτο κατὰ πτύχας Οὐλύμποιο."—'Il.,' xi. 76.

sacred character to the highest summit in a country;[17] so that there is no need, with Niebuhr, to suppose that it must once have been situated in the midst of the great Pelasgian nation.[18]

In describing the general view I have mentioned some lofty summits of Olympus which rose to the south-west of St. Elias. The nearest of them, which was quite precipitous on the side that faced us, with a considerable mass of snow at its foot, was separated from us only by a deep ravine, and being broad at its base, serves in great measure to bar all access to the part of the mountain beyond it. M. Heuzey with his usual inaccuracy, speaks of St. Elias as being the highest summit of Olympus; but this was certainly higher, and behind it again appeared another peak, rising to a somewhat greater elevation still. This one both my companion and myself estimated as being about 200 feet above the point on which we were standing. The barometer gave the height of the summit of St. Elias as 9758 feet, and the coincidence here is sufficiently striking with 9754 feet, given as the height of Olympus by the trigonometrical survey for the English Admiralty chart: but if this latter represents the real summit, it seems to imply that the difference between that and St. Elias is not very great. Dr. Barth, who approached Olympus from the west, describes a group of three peaks, in which St. Elias is not included, and the central one of which rises above the others, as forming a very conspicuous object from that side, and being evidently the highest points of the mountain.[19] I

---

[17] *E.g.*, Mount Meru among the Indians, Mount Elbrouz among the Persians. *See* Preller, 'Griechische Mythologie,' i. 50, *note*.

[18] 'Lectures on Ethnography and Geography,' i. p. 287.

[19] 'Reise,' p. 186. Barth, who passed close under these peaks in

have little doubt that the two which we saw were two of these, the third being hidden behind the others; and at the monastery, on our return, we were told that these three summits bear the name of "the three brothers" (τὰ τρία ἀδέρφια). From a song, which is found with slight variations in various parts of Greece, they seem to have a mysterious importance attached to them, and to be called "the three peaks of Heaven." The song runs as follows :—

> "From the summit of Olympus,
> From the triple peak of Heaven,
> Where the Fates of Fates abide,
> May my fated destiny
> Hear my call, and visit me!"[20]

We were already aware, from Dr. Barth's account, that crossing the saddle from the western side of the mountain into the central basin, seems to have greatly under-estimated the height of St. Elias relatively to them, when he speaks of "die um mehrere Tausend Fuss niedrigere Vorkuppe des heiligen Elias" (p. 194). If this were so, it would be almost impossible that in the view of the mountain from Katrin—in which they are foreshortened, and appear as one peak—St. Elias should appear the highest, even allowing for its being somewhat nearer; nor does the appearance of these summits from the valley, nor Barth's own view, taken from the western side (p. 164), give the idea of their rising to so great an elevation above the ridge of the mountain as to render such a difference conceivable.

[20] "'Απὸ τὸν Ὄλυμπον τὸν κόρυμβον,
τὰ τρία ἄκρα τοῦ Οὐρανοῦ,
ὅπου αἱ Μοῖραι τῶν Μοιρῶν,
καὶ ἡ ἐδική μου Μοῖρα
ἂς ἀκούσῃ καὶ ἂς ἔλθῃ."

This is given by Heuzey, 'Le Mont Olympe,' p. 139. Other versions are found in Wordsworth's 'Athens and Attica,' p. 231, and in No. 574 *b* of Passow's collection, which is derived from Ulrichs. There seems but little doubt that the word Μοῖρα, which sounds so enigmatical, here has especial reference either to marriage or to deliverance in childbirth. As to the former, in the Romaic ballads καλὴ μοῖρα νὰ λάβῃς signifies, when addressed to a woman, "may you get a husband;" κακὴ μοῖρα νὰ λάβῃς, the contrary (Passow, Nos. 442, l. 7, and 539, l. 6); and in a song which it is customary to sing while the bride is being adorned for her wedding, the

the peak of St. Elias was not the highest, and on leaving the monastery in the morning our intention had been, after mounting to that point, and reconnoitring from it, to pass the night either in the chapel or at the nearest place of shelter below, and the next day to attack the highest peak, which had never yet been attempted. The only other ascents of Olympus which I know of besides our own are those of Heuzey and Barth, both of whom reached St. Elias: Mr. Urquhart also has described an ascent in his 'Spirit of the East,' but he only mounted one of the lower peaks of the southern range. Our prospect of ascending further certainly did not appear a very hopeful one. The side of the nearer of the two high summits which we could see was absolutely precipitous, and the cliffs that descended from it into the great basin, were steep enough to make it an arduous task to get round them to the base of the central peak. The ascent of this would then remain, and as we were without guides, and destitute of the ordinary appliances of mountain climbing, the chances of ultimate success seemed very doubtful, for it had the appearance of a regular *aiguille*. However, we went to confer with our attendants, who had taken shelter under a rock, and were shivering in their thick capotes; but when we proposed to them to pass the

---

mention of this triple peak again occurs (Wachsmuth, 'Das alte Griechenland im neuen,' p. 88). As regards childbirth, it is the custom even to the present day among the Athenian women, when they are with child, to slide down a place on the side of the Hill of the Nymphs, singing at the same time "ἔλατε μοῖραι τῶν μοιρῶν, νὰ μοίρατε κ' ἐμένα" (Wachsmuth, p. 71, and Pouqueville, as there quoted). Dr. Wordsworth also tells us that the song given above is sung in a grotto at Cephissia, nine miles from Athens, by the female peasants who come there to inquire their future destiny. It should be remembered that in classical times the Μοῖραι presided in an especial manner over marriage and childbirth, so that Aphrodite Urania was sometimes regarded as one of them, and Eileithyia is spoken of as their associate (Preller, 'Griechische Mythologie,' i. 414).

night in the chapel, they at once exclaimed, "We shall die here;" and indeed, though the thermometer in a sheltered spot had not gone down below 46°, yet in the wind it was bitterly cold. Accordingly we descended to the plateau at the foot of St. Elias, and being unwilling to relinquish all hope of carrying out our plan, were still entertaining the possibility of sleeping at the shepherd's *mandra*, which could not have been far from the place where we met him with his goats, when the matter was settled for us by Jove himself in a very summary manner. Before leaving the peak we had observed a storm rapidly approaching from the west, attracted apparently by the lofty summits; and whilst we were engaged in making a hurried meal on the edge of the snow-basin, which supplied us in a solid form with the only drinkable, except brandy, that could be procured, it burst upon us in violent wind, sleet, and snow, before which there was nothing to be done but to fly. To return along the *arête* seemed hopeless, so we clambered down a steep gully or *cheminée* in the rocks which here overhang like a crust the central basin, and commenced zigzagging down its side, as steep and rough a place as an Alpine climber could desire. The story of our descent is soon told, but took a long time to accomplish, for none of our party had ever explored that part of the mountain, and we were left to conjecture as to the best way of reaching the valley. Our young Greek, however, from his intelligence and presence of mind, here proved highly serviceable. Unfortunately, all botanizing was now at an end; but this was the less to be regretted, as the higher parts of Olympus, owing to the bareness of the rocks and the stony *débris*, are singularly destitute of flowers, and the only ones of much interest which I had found during our ascent were some fine specimens of a yellow and a purple *anthyllis*.

At the end of a long scramble we arrived at the region of trees, where our difficulties rather increased than otherwise, as the mountain-side was hardly less steep, and the fallen trees and rotten bark and branches made the way treacherous and slippery. At first we followed a watercourse, along the sides of which we had to climb, wherever there had been cascades in the winter, or where the snow was lying, for it reached even as low as this part: afterwards, having to turn a ravine which we found impracticable, we were forced to penetrate further into the wood, which is called Μαυρολόγγο, or the Black Forest. In the midst of this, night came on; and as the storm was now over, and the sky perfectly clear, we began to think of camping out, when our lively Palicar, who had long shown signs of distress, plumped himself down on the grass, and declared he would go no further. This act of insubordination at once determined us to proceed, and at last we saw a light shining far below us, to which, after many falls, we succeeded in making our way. It proved to proceed from a shepherd's encampment hard by a small stream, in which we slaked our thirst, for we had had nothing but snow to drink since the middle of the day. The moon had now risen, and the view of the stupendous cliffs and pointed summits all round us was indescribably sublime. From this point a track sufficiently well-marked conducted us to a saw-mill in the bottom of the valley, the inmates of which undertook to give us lodging for the night. As we were eating our supper, and looking forward to a bed of sawdust and shavings, we happened to enquire what distance we were from the monastery, and were told that it was only half an hour off. "Oh! no, it's at least an hour-and-half," exclaimed our weary Palicars. "That's all lies!" we replied, "all lies!" using the ordinary polite Romaic expression of incredulity[21] (ὅλα ψέμματα), which was followed by a general laugh, in which the Palicars themselves joined. Now, there is a curious proverb among the Modern Greeks, aimed at those who give in when on the point of accomplishing an undertaking, which runs as follows:—

> "We ate the ox, and finished him;
> The tail we could not manage."[22]

For our part we had no intention of rendering ourselves amenable to such an accusation; so shortly afterwards we were once more *en route*, the whole party having regained their spirits, and following a path down the valley, which after our previous scrambling seemed to us a royal highway, an hour before midnight we reached our destination.

---

[21] Wachsmuth remarks that in every part of Greece, except Acarnania, the expression Ψεύματα λέγεις is universally regarded only as a form of compliment ('Das alte Griechenland im neuen,' p. 46).

[22] "ὅλο τὸ βώδι τὸ φάγαμεν,
καὶ 'ς τὴν οὐρὰν ἀποστάσαμεν."—Ross, 'Inselreisen,' ii. 176.

CHAPTER XIX.

THE LOWER OLYMPUS.

Letochoro — District of Pieria — Leftocarya — Monastery of Kanalia — Monastic Contentment — Western Side of Olympus — Monastery of Sparmos — Suspicions of the Monks — Lake of Nezero — Mixture of Races on Olympus — View of the Plain of Larissa — Battle of the Gods and Titans — Clefts and Armatoles — Their Origin and History — Encroachments of the Turks — Clefts of Olympus — Popularity of the Clefts — Vigorous Measures of Ali Pasha — National Movement — Disbanding of the Armatoles.

THE following morning was bright, but the distance was once more obscured by haze, and clouds rested on the summits of the mountain; so we considered ourselves fortunate in having accomplished our ascent so well. In the afternoon we returned to the *Scala*, and proceeded from thence to the village of Letochoro, which is only half an hour's ride distant. Just before entering it we passed through a deep channel with precipitous banks, at least 70 feet below the ground at its sides, which has here been worn away by the stream that descends from the monastery, and in winter must be swollen to a furious torrent.[1] The view of the summits of Olympus, as seen from this point through the steep cliffs at the opening of the gorge to the west of Letochoro, is truly magnificent.

[1] This must be the Enipeus of ancient writers, for that river is stated by Livy to be five miles on this side of Dium, and this corresponds very closely with the distance from the ruins at Melathria (Livy, xxiv. 8). The historian gives the following characteristic description of it:—"It flows from a valley in Mount Olympus, and in summer is a narrow stream; but when swelled by the winter rains it eddies along in a mighty current over the rocks, and by rolling towards the sea the soil which it has swept away, forms whirlpools of great depth in its lower course, with steep banks on both sides, from the channel between them being excavated."

The deep valley which reaches from the saw-mills above St. Dionysius to the exit of the stream at this place is called, like several similar valleys in Olympus, by the name of *lakkos*.[2] The village is the head-quarters of the manufacture of *skutia*, or rough cloth for capotes, and has been more fortunate than the villages on the west of Olympus, which I have already spoken of as being ruined by English competition, for here the trade flourishes as well as ever. The rest of the population who are not employed in this way are mostly sailors; they are all Greeks, and seem to form a very thriving community. As usual in flourishing Greek villages, they have an excellent school. A priest who is at the head of it paid us a visit, and from him we learned that it numbers 250 pupils, all of whom are taught to read and write, while the higher classes read Thucydides and Demosthenes, and some even learn Latin. We had numerous other visitors in consequence of the fame of our ascent, of which a flaming account had no doubt been circulated by the Palicars, whom we had sent off early in the morning: they were very anxious to hear all about it, and could hardly believe that snow had fallen on the summit. What struck us most forcibly about these people was their acuteness and inquisitiveness, which formed the most complete contrast to the dulness of the Turks and Bulgarians, amongst whom we had so long been travelling. The aneroid especially interested them, and its power of measuring altitudes; and in calculating the height of a mountain and comparing it with other mountains, they could appreciate differences of 100 feet, or even smaller numbers, a thing which a Turk would never think of.

---

[2] The same name is applied to Messenia, as being a deep-sunk valley-plain (Wyse's 'Excursion in the Peloponnesus,' i. p. 258), and is said to be the derivation of "Laconia."

The next day we pursued our way southwards along the lower slopes just below the spring of the steeper flanks of the mountain, with views of the castle of Platamona in front of us, on a height close to the sea, conspicuous from its white walls and tower, with the peak of Ossa rising behind. Platamona was the site of the city of Heracleium, and it must always have been an important position, as it commanded the entrance of the plain of Macedonia from the pass of Tempe. After crossing in succession seven river-beds similar to that at Letochoro, and overgrown with plane-trees, we arrived in two hours at the village of Leftocarya, which is situated on a hill-side near one of the deepest of these water-courses.[3] This place

[3] It is probable enough, as Heuzey suggests ('Le Mont Olympe,' pp. 96, 112), that the river here is the Sys of ancient times, which is reported by Pausanias ('Bœotica,' xxx. 5) to have swept away the city of Libethrium, in consequence of the neglect shown by its inhabitants to the tomb of Orpheus. Now, from Livy's account of the campaign of the Consul Marcius against Perseus, it evidently appears that Libethrium lay between Heracleium and Dium, because when entering Pieria, on his way to Dium from the southern district of Olympus, he is said to have descended to the lower country at a point between Heracleium and Libethrium. Leake ('Northern Greece,' iii. 422) is disposed to place the city further to the north, and to identify the Sys with the Enipeus, as he thinks the river of Letochoro the only one in this neighbourhood which could effect the ravages which Pausanias describes. But the stream of Leftocarya when swollen would become a great torrent, and it is not likely that one stream should have been called by two names; accordingly it seems more reasonable to place Libethrium some way below this village towards the shore, where, from the country being more level, a great inundation would be possible, which would not be the case in its upper course, where it flows between steep banks. The name of the river—the Boar—is well adapted to express the violence of these mountain torrents, that animal being constantly introduced in Homeric battle-pieces as a simile for a sudden rush or attack. If we are right in the position to be assigned to Libethrium, then Leftocarya may very well represent Pimpleia, or Pimpla, which is mentioned by Strabo (x. i. § 17) in connexion with that place, and was consecrated to the worship of the same deities. Apollonius also speaks of its commanding position ("σκοπιῆς Πιμπληΐδος."—Ap. Rhod., i. 23), which well agrees with the wide sea view obtained from this point. Higher up than this the mountain sides are too abrupt to allow of the site of a city.

was probably the ancient Pimpleia, which was famed as being the birthplace of Orpheus, the father of song; here he is said to have established the worship of Dionysus and his mysteries. The rest of the district of Pieria, too, which we have lately been traversing, is connected with many of the earliest and most romantic classical associations, for here the Muses were born, and from it were derived many of their most familiar names—Pierides, Libethrides, Pimpleides—which were afterwards transferred, together with their worship, to the sides of Helicon.

At Leftocarya we turned directly westward, and began once more to penetrate into the interior of the mountain. We were now entering the lower Olympus, that is, the part which extends from the outer flanks of the high southern range that has been already mentioned as bounding the central basin, to the vale of Tempe. Though inferior in altitude to the northern part, it is still a very elevated district, composed of lines of mountains which for the most part run parallel to the loftier chains, opening at intervals to leave room for small upland plains and valleys. The descent from it to the surrounding country is on every side extremely steep. Its geological character, too, is different from that which we have hitherto explored; both in the neighbourhood of Leftocarya and throughout a large part of the Lower Olympus the limestone is replaced by igneous and metamorphic formations, and the character of the scenery, as we shall soon notice, undergoes a corresponding change. We ascended by zigzags at the side of a deep valley which runs into the heart of the mountain, through delightfully aromatic undergrowths: a little way above the village are found the hazels from which it takes its name, though there are none in the place itself; and on the lower slopes the Spanish chestnut abounds. It is noticeable that though the cypress is found everywhere on Mount Athos, there is not one in any part of Olympus. On reaching the summit of the long ascent we struck inwards by a winding path along the sides of the mountains, and after three hours reached the little monastery of Kanalia, which overlooks the valley, or chasm, as from its precipitous character it might almost be called. It is from this feature,—and not, as I had expected, from the pretty rivulets and watercourses by which a supply of water is brought into the building,—that the name is derived, for the word Kanalia, according to its inhabitants, is used in the same way as *lakkos*, for a narrow gorge; and it is found again in the neighbourhood of Lake Bœbe in Thessaly, applied to a village in a similar position.[4] Southward from this point a high peak, which is here called Durjani, appeared above the nearer range; it is said to be near the lake of Nezero, to which there is an easy pass from Kanalia.

This convent contains four monks, together with thirty seculars; it has no objects of interest to show, and the buildings are of a very rough description. Their revenues are derived from their flocks and from a little woodcutting; also from the number of mares and foals, which we saw feeding on the mountain sides, they seem to breed horses. One of the monks, however, notwithstanding the rudeness of the place and its retired situation, was a singularly intelligent man. He talked much of the two visits of Prince Alfred and Prince Arthur to the Vale of Tempe, which appeared to have made considerable impression in these parts, and remarked of the former, "he ought to have been King of Greece." He referred, also, to the Queen's retirement from public life, and understood the distinction between our Upper

[4] Leake, 'Northern Greece,' iv. 421.

and Lower Houses of Parliament, but he was firmly impressed with the idea that the English were the great supporters of the Turks. He spoke familiarly of Lord Russell, Lord Palmerston, and Mr. Gladstone; and, what surprised us most of all, he was aware that Lord Derby (ὁ Λόρδος Δέρβυ) had published a translation of 'Homer'—an unexpected tribute of praise to the noble author, that the fame of his work should have spread even to Jove's own mountain. Like all the monks in these parts he expressed his warm admiration of the greatness and splendour of the monasteries of Athos; and, indeed, it was not until we had seen something of these very simple establishments and their humble occupants, that we were able to appreciate the idea that is entertained of Athos throughout the Eastern Church. As we looked back upon them from this lower level, the size and grandeur of the buildings, the wealth and antiquity of the institutions, the works of art they contained, and the courtesy and intelligence of their inhabitants, appeared altogether imposing. Even the good Romaic which they spake, when compared with the rude jargon of these parts, made us picture them to ourselves as quite educated men; and we were not surprised to find that the caloyers here fancy that they are acquainted with many languages.

I could not help feeling how great the temptation must be for so intelligent a man to betake himself to Athos, where he would enjoy a wider range of ideas and more means of communication with the outer world, and where in all probability he would soon rise to a high position. When I put this before him he at once answered that he felt no such desire; that prayer was the monk's occupation, and while they had that they were content. There was something in his reply and in his general deportment which I have remarked in other Greek monks, and which, though mixed no doubt with an element of torpor, seemed to breathe the spirit of those divine lines of Dante, in which he solves the doubt whether those who attain to the lower mansions of bliss in heaven will be satisfied with their lot :—

> " Frate, la nostra volonta quieta
>   Virtù di carità, che fa volerne
>   Sol quel ch' avemo, e d' altro non ci asseta.
>
> Se disiassimo esser più superne,
>   Foran discordi li nostri disiri
>   Dal voler di colui che qui ne cerne,
>
> Che vedrai non capere in questi giri,
>   S' essere in caritate è qui necesse,
>   E se la sua natura ben rimiri;
>
> Anzi è formale ad esto beato esse
>   Tenersi dentro alla divina voglia,
>   Perch' una fansi nostre voglie stesse.
>
> Sì che, come noi sem di soglia in soglia
>   Per questo regno, a tutto il regno piace,
>   Com' allo re ch' a suo voler ne invoglia;
>
> E la sua voluntade è nostra pace;
>   Ella e quel mare al qual tutto si muove
>   Ciò ch' ella cria e che natura face."
>
> *Paradiso*, Canto iii.

> " Love by his virtue, Brother, hath appeased
>   Our several wills: he causeth us to will
>   But what we have, all other longings eased.
>
> Did we desire a region loftier still,
>   Such our desire were dissonant from His,
>   Who bade us each our several station fill:
>
> A thing impossible in these spheres of bliss,
>   If whoso dwelleth here, in Love alone
>   Must dwell, and if Love's nature well thou wis.
>
> Within the will divine to set our own
>   Is of the essence of this Being bless'd,
>   For that our wills to one with His be grown.

> So, as we stand throughout the realms of rest,
> From stage to stage, our pleasure is the King's,
> Whose will our will informs, by Him imprest.
>
> In His will is our peace. To this all things
> By Him created, or by Nature made,
> As to a central sea, self-motion brings."
>
> Gladstone's Translation.

Beyond Kanalia the character of the scenery changes, for the vegetation becomes scanty, and the country generally assumes a dreary aspect. The valley continues to penetrate westward into the heart of the mountain; but about an hour's distance from the monastery it opens out into a narrow plain about five miles in length, bounded by the same two parallel ranges which form the valley; and as the water-shed is at its further extremity, the stream that flows through it intersects the whole breadth of Olympus, while the descent towards the lower ground on the western side is consequently extremely rapid. In this direction a strange knobbed peak appears, which was called to us the mountain of Elassona, from the town of that name which lies on its further side; it is conspicuous from everywhere in this neighbourhood, and presents the most marked point in a succession of summits descending from east to west, which forms a continuation of the range that rises to the south of this upland plain. The appearance of these summits, independently of their position, would suggest that they represent the mountain called Octolophus, or the "ridge with eight crests," in a passage of Livy to which we shall hereafter have to refer. In a recess of the mountains on the north side of the plain lies the village of Carya; above this a steep ascent leads to the summit of a pass, the view from which, extending far away towards the west, formed a most satisfactory supplement to our panorama from the peak of St. Elias. Below us lay the undulating land in which the Sarandoporos, the ancient Titaresius, flows; beyond this rose the conspicuous form of Mount Amarbes, the highest point of the Cambunian chain, while far in the distance the superb range of Pindus bounded the view with a succession of broken summits. To the north appeared the rounded grassy heights, which lead up to the lower St. Elias, the highest point on the southern side of the great central basin: they presented a marked contrast to that part of Olympus which we had ascended, from the entire absence both of sharp outlines and of the delicately cut forms of the "many-folded" buttresses. As I have already said, this difference is owing to the variety in their geological formation. We descended on the other side of the pass to Scamnia—a good-looking village with a tall bell-tower—where we hired a guide to direct us to our destination, the monastery of Sparmos, as the evening was beginning to close in. After two hours of very rough riding along the bare mountain side, we reached this remote spot just before the last glimmer of daylight had disappeared. So completely had we turned the south-west angle of Olympus that, during the latter part of this journey, our direction had for some time been due north.

Our arrival was greeted by a loud barking of dogs, who were keeping guard outside the building. Then followed a long parley at the gate, for the hegumen was absent, and the monks were extremely unwilling to admit us, so that it required all the influence of our local guide to persuade them to give us a lodging. The whole affair reminded us forcibly of the scene so amusingly described, as happening at the same place, by Mr. Urquhart in his 'Spirit of the East.' When at last we were admitted we found our hosts a very rustic set—so much so that it was almost worth while coming so far to learn how extremely rough some of these caloyers are. As a natural consequence their suspicions knew no bounds. They could not comprehend our object in visiting so many monasteries, and thought we must have something sinister in view : in particular they greatly disliked the idea of having drawings taken of their buildings, being probably possessed by the notion common among the lower classes that, together with the likeness, some property belonging to the thing or person sketched is taken away, or some power obtained over them. One cause of their fear seemed to be that two Englishmen (so their story ran) had been somewhere in the neighbourhood a few years before, with two chests in their possession, containing so deadly a poison, that one drachm of it sufficed to kill several hundred people. What could possibly be the ground of this extravagant notion we could not discover; but fancies of this kind have always had a strange hold on the imagination of the vulgar ; as witness the account given by Manzoni, in his 'Promessi Sposi,' of the ideas which were abroad on this subject at the time of the great plague in Milan, and which were shown not to be extinct by their reappearance in Italy during the cholera in 1866. As to the age of their monastery, or the name of the founder, they were wholly ignorant ; and to every question on such subjects they invariably answered "Nobody knows! nobody knows!" (ποῖος τὸ ἐξεῦρε;)

The name of the convent is not Hagia Triada, as Leake and Heuzey[5] state, for that name is neither more nor less applicable to it than to the other monasteries

---

[5] Leake, 'Northern Greece,' iii. 349; Heuzey, 'Le Mont Olympe,' 52.

on Olympus, which are all dedicated to the Holy Trinity—but Sparmos, *i. e.*, 'Cornhill;' this being derived from the sloping plain a little way below it, almost the only spot in the neighbourhood that admits of cultivation. The corresponding, though more famous names of Sparta, "the sown land," and Jezreel, "the sowing place," are similarly attached to positions somewhat elevated, overlooking fertile plains. Its situation is very striking, being in the midst of trees at the side of a narrow gorge, on a shelf of land which lies under the huge western buttresses of Olympus, at a height greater even than that of St. Dionysius above the sea. But the buildings are equally rough with those of Kanalia; there is no attempt at architecture even in the church, nor any Byzantine work, such as is usually found in these edifices. The frescoes on the interior walls of the church and a few of the pictures were somewhat better than usual, but the only objects really deserving of notice were the wood-work of the Iconostase and bishop's seat (they are under the bishop of Livadi), which are boldly and richly carved in birds and flowers. In former days, when Olympus was one of the great head-quarters of the Clefts (κλεφτοχώρια), Sparmos was known as one of their favourite haunts. In the ballads which relate to these mountains, the wounded Cleft is usually represented as sending for help to one of the monasteries.[6]

We had accomplished our main object in visiting this place, which was to get a clear idea of the western side of Olympus and the neighbouring country. If time had been no object, it might have been worth our while to descend into the district to the north-west of this point, where are a number of important villages occupied by

---

[6] *E.g.*, No. 154 in Passow's collection.

a Wallach colony at the foot of the mountain. This, however, would have led us too far out of our intended course, which lay in the direction of Tempe; and accordingly the next morning we retraced our steps as far as the village and plain of Carya. During the best part of the way the ridge of Octólophus formed a conspicuous object before us, bounding the view; between this and Scamnia rose a mountain called Boliano, with a village of the same name on its side, below which the ground descends steeply towards the lower land, which it continues to do all the way to Sparmos. Above Scamnia is another height, called, by Heuzey, Detnata, over the eastern shoulder of which we had passed from Carya on the previous day: in returning we crossed its western flank by another track, thereby obtaining additional views of the surrounding objects. When we had reached the foot of the mountains on the southern side of the plain of Carya, we mounted by a long and steep ascent to the summit of the pass leading over them, about 4020 feet above the sea, looking back from which we obtained a splendid view of the southern range of the higher Olympus, while in the opposite direction the elegant grey peak of Ossa appeared in the distance, and at our feet the Lake of Nezero deeply embedded in its cradle of mountains—a glassy expanse of blue water some five miles in circumference, with green marshes and some corn land about its head. The rushy islets with which its surface was dotted recalled to my mind the curious lake with floating islands which Pliny has described in one of his Letters. The valley in which it lies is about three miles in diameter, and nearly circular, leaving no visible means of escape for its waters; the general direction of the lake is from north-east to south-west, and at no great distance from its head is situated the village of

LAKE OF NEZERO MOUNT OSSA IN THE DISTANCE.

Nezero, to which we descended for the night. At this its northern end there are three mountains of considerable elevation, the central one of which, lying between the pass by which we crossed and the village, is called like so many others St. Elias, but also bears the name of Chouka, which is said to be a Bulgarian word for "summit." To the east of this rises Mount Livadaki, the highest point in the lower Olympus, which is called, by Heuzey, Metamorphosis, or the peak of the Transfiguration, and is the same which the monks of Kanalia named Durjani. Between these two mountains is the pass which leads to that monastery; and from here it appears low and easy, though from the relative elevation of the ground the descent must be considerable on the other side. On the southern side of Livadaki, and east of the lake, there is an opening of some width, through which a track leads to the village of Rapsani, on the heights which overlook the vale of Tempe. The route from Sparmos to that place was in all probability the direction followed by the Roman Consul Marcius, who led an army across Mount Olympus in order to force his way into Macedonia in the war with Perseus.[7]

The village of Nezero, though situated at some little distance from the lake, is infested with mosquitoes, and a most unhealthy place, as is sufficiently shown by the haggard looks of its inhabitants. In consequence of this, the project was long ago started of draining the whole valley by means of a tunnel bored through the mountain side, similar to that by which the Alban lake was carried off. Colonel Leake makes mention of an attempt of this kind, and last year it was renewed by the Mudir of the district; but after the work had proceeded for a month,

[7] See Appendix F, "On the March of a Roman Consul across Mount Olympus."

the money failed, and it was left unfinished. The mouth of the tunnel is visible at the south-west end of the lake. The name of the village is peculiar in its form, being derived from Ezero, the Bulgarian word for a lake, together with the prefix *n*, which is the termination of the accusative case of the Greek article attached to the noun. Similar instances are found in Nisvoro, the modern form of the ancient Isboros, Negropont from Egripo, the corruption of Euripus, the full form having been ἐς τὸν Ἔζερον, ἐς τὸν Ἴσβορον, &c.; this is found almost complete in Stalimene (ἐς τὴν Λῆμνον), Stanco, (ἐς τὴν Κῶ), the modern names of Lemnos and Cos. Again, in plural names the *s* of the article became prefixed, as in Satinas (ἐς τὰς Ἀθήνας), the ordinary name for Athens in the time of Spon and Wheler, while here again the full form may be seen in στοὺς στύλους, the peasants' name for the remains of the Temple at Bassæ, in Arcadia.[8] A corruption of a corresponding character may be traced in some modern words, such as the English "newt" for "an eft," and the German "natter" from the older form "otter," which is the same as our "adder," in old English "nadder." In fact, though changes produced by this kind of agglutination can be reduced to no rule, they are not uncommon in the decay of languages, especially in proper names. But in the case of Nezero there is a still further peculiarity, for the component parts of the word proceed from different languages, and testify to a Bulgarian population being either themselves *hellenized*, or else overspread or succeeded by a Greek one. This mixture of races on Olympus is not the least noticeable feature of the district. We may apply to it the remark which

---

[8] Leake's 'Travels in the Morea,' ii. 1.

has been made with regard to the Peloponnese in classical times, that from its rugged character and isolated position it retained remnants of the different races which had occupied it from time to time. The prevalence of Slavonic names in various localities, such as Golevo, Chouka, and (to take a marked instance) Tzaritzena, *i.e.*, the Royal village,[9] situated at its south-west angle, testify to the prevalence of some branches of that race at one period. An instance similar to that of Nezero, where a Greek and a Slavonic element are found in combination, is Mikro Gurna, or "the smaller ravine," the name of a rocky valley on the western side of the highest summits, so called to distinguish it from Trani Gurna, "the great ravine," where both words are Slavonic.[10] The river Sarandoporos in addition to this Greek name is also called Vurgaris, "the Bulgarian."[11] And while the population of Letochoro is thoroughly Greek, the physiognomy of the people of Nezero, though they speak Romaic, greatly resembles the Bulgarian type. The struggle between these two nationalities was still in progress when, shortly before the taking of Constantinople the Koniarates, a colony of Turks from Asia Minor, were planted under Olympus towards the plains of Thessaly: with these we shall meet as soon as we descend to the foot of the mountain. The Wallachians again, who at one period of the middle ages so completely occupied Thessaly, that that country was called by Byzantine writers Great Vlakhia, or Wallachia (Μεγάλη Βλαχία), and who were found by Benjamin of Tudela as far south as the Maliac gulf, have left their representatives in the ancient colony still inhabiting Vlacho-Livadi and other villages in the neighbourhood of the pass of

---

[9] Heuzey, p. 21.   [10] Barth, 'Reise,' p. 188.
[11] Leake, 'Northern Greece,' iii. 334.

Petra. To these, if we wished to swell our enumeration of races, we might add the Albanians, from which nation the guards of the mountain have for a long time been drawn.

We continued our journey along the lake until we reached the foot of the mountains at its southern end, and when we had mounted to the summit of the pass, which is about the same height as that by which we had entered the valley, looked back on the long line of rounded summits which form the southern chain of Olympus, culminating in the high peak of St. Elias. As we descended on the other side, there opened out before us to the south and south-west a grand view of the great plain of Larissa, which forms one of several sections opening out into one another, into which the plain of Thessaly is broken up. A long spur is here seen to run down from the Cambunian mountains, reaching nearly to the city of Larissa, the position of which in the midst of its gardens was pointed out to us, though we could hardly distinguish more than a dim spot owing to the haze. Behind it were other mountains, and to the south the Karadagh, or Black Mountain, on the nearer side of Pharsalia, one of the innumerable ranges in Turkey that bear that name. The course of the Peneius was visible as it wound through the plain; and in one place a line of green marshes appeared, marking the overflow of its waters from the right bank during the floods of spring, by which the lake Nessonis is formed. That piece of water was a conspicuous object as it shone in the sun beneath the flanks of Ossa, and seemed to the eye to be of considerable size; when it becomes full, the surplus water descends into the lake of Bœbe, which, however, is here concealed from view. To the eastward lay the valley of the Titaresius, and the plain of Turnavo, though the town of that name is hidden by an intervening spur.

With the great Olympus behind us, and in front this vast expanse, beyond which in clearer weather the long line of Othrys would appear to the south, bounding the horizon, it was a fine position from which to conceive the battle of the Gods and Titans, as Hesiod has described it in the magnificent passage from which the following is an extract. After assigning to the combatants their respective positions, to the Gods the heights of Olympus, to the Titans those of Othrys, he thus represents the final struggle :—

> " Th' immeasurable sea tremendous dash'd
> With roaring; earth re-echoed; the broad heaven
> Groan'd shattering: vast Olympus reel'd throughout
> Down to its rooted base beneath the rush
> Of those immortals: the dark chasm of hell
> Was shaken with the trembling, with the tramp
> Of hollow footsteps and strong battle-strokes,
> And measureless uproar of wild pursuit.
> So they against each other through the air
> Hurl'd intermix'd their weapons, scattering groans
> Where'er they fell. The voice of armies rose
> With rallying shout through the starr'd firmament,
> And with a mighty war-cry both the hosts,
> Encountering, clos'd. Nor longer then did Jove
> Curb down his force; but sudden in his soul
> There grew dilated strength, and it was fill'd
> With his omnipotence: his whole of might
> Broke from him, and the godhead rush'd abroad.
> The vaulted sky, the Mount Olympus, flash'd
> With his continual presence; for he pass'd
> Incessant forth, and lightened where he trod.
> Thrown from his nervous grasp the lightning flew
> Reiterated swift; the whirling flash
> Cast sacred splendour, and the thunderbolt
> Fell. Then on every side the foodful earth
> Roar'd in the burning flame, and far and near
> The trackless depth of forests crash'd with fire." [12]

---

[12] Hesiod, 'Theog.', 678-694 (Elton's Translation).

It was not, however, of the wars of Gods and Titans that we were talking as we looked down from this eminence, but of a much more real event, the Olympian insurrection of 1854, when the inhabitants of the mountain rose to assist the Greeks of the south in their ill-advised raid into Thessaly. The guide, whom we had hired at Nezero for fear of losing our way on these unfrequented mountain sides, had taken part in this rising, and described to us how they had collected 500 men at this point, and were within one day of taking Larissa, which, in fact, they would have done, had not the Turks been able to call up reinforcements just in time to save the place. His narrative so strikingly recalled the old days of the Clefts and Armatoles; and the strange history of these bands, and of the captains of Olympus, supplies to the modern Greeks so much of what the tales of heroes and demigods offered to their heathen forefathers, that it may be worth while here to give a slight sketch of the former condition of things which has now completely passed away, and to explain the circumstances by which it was brought about.[13]

To begin with the Armatoles. This name, which signifies an armed man, or man-at-arms (Ἁρματωλός, or, as it is often written by a curious transposition of the letters, Ἁμαρτωλός), was applied to a local militia, or rural police, composed entirely of native Christians, to whom was entrusted in great measure the security of the country where they were established. The institution extended over the greater part of those provinces of Turkey which were occupied by a Greek population,

---

[13] The completest account of these is that given in the Introduction to Fauriel's 'Chants Populaires de la Grèce,' pp. xlii.-lxxix. The greater part of the following notice has been derived from that source.

from the banks of the Vardar, towards the north, to the gulf of Corinth to the south, including also Epirus, Ætolia, and Acarnania. These were divided into districts, varying in number at different times, in each of which a separate body of Armatoles was organised, with a captain of its own, whose office was hereditary. The individuals who composed these bodies commonly called themselves Palicars, and wore the dress and arms of an Albanian soldier; the captain was styled Protopalicar. In those parts of the country which were governed by a Pasha, they were supposed to act under his orders; but where there was only a governor of inferior rank, their services were at the disposal of the local Greek authorities, the Primates. History is altogether silent as to the origin of this system, and the date of its establishment; but the completeness and uniformity of its organisation seem to show that it did not grow up by accident, but was the development of a definite plan: and while on the one hand there is no trace of its existence before the conquest of these lands by the Turks,[14] on the other the fact that it never extended to the Morea, seems to suggest that it was established before the Turkish conquest of that country. In addition to this negative evidence, there are traditions existing among the Greeks on the subject, which furnish us with a very probable explanation of its rise. According to these, when the conquerors established themselves successively in the different provinces of Greece, they found the population of the plains tractable enough, and ready to submit to their yoke; but with the inhabitants of the mountains the case was different, especially in the neighbourhood of

---

[14] Mr. Finlay says ('History of the Greek Revolution,' i. p. 24) that a Christian local militia had existed in the Byzantine Empire, but he does not give us his authorities.

Thessaly, the first province that fell into their hands. Olympus, Pelion, and Pindus at first, and afterwards the mountains called Agrafa, under which name are comprised a number of confused mountain chains in Acarnania and the west of Thessaly, became the headquarters of a permanent resistance to the invaders; and the strength of these rocky fastnesses, and the warlike spirit of the mountaineers, rendered a continued struggle with them at once vexatious and unprofitable. Accordingly, the Turks found it the wisest plan to come to terms with them, and employ them as far as possible in their own service. With this view they offered them favourable conditions in case of their submission; and whilst some of the most resolute refused to treat, and, retiring to inaccessible positions, established their headquarters in the Cleft villages, as they were called ($\kappa\lambda\epsilon\phi\tau o\chi\omega\rho\iota a$), the greater number recognised the authority of the conquerors on the understanding that their rights should be respected, and that they should be allowed to form a militia among themselves for the maintenance of these rights, as well as for the preservation of order throughout the country. This was the institution of the Armatoles.

The advantages of this arrangement were great to both parties. To the Greeks it guaranteed a certain amount of freedom from interference, and security from the violence of the Turkish soldiery, and in some cases even the possession of their lands. To the Turks, on the other hand,—who from their remaining distinct in race and creed from those among whom they lived, were hardly more than settlers in the country,—it was an easy means of preserving the goodwill of the subject race, and securing that tranquillity on which the payment of tribute and the safe possession of their conquests depended. But the Turkish authorities were not satisfied with this; their aim was to establish their domination more completely, and to despoil the Christian population more thoroughly. The Armatoles were an obstacle in the way of their accomplishing this, and consequently their history is the narrative of a long struggle for independence against the attempts of the Pashas to compass their overthrow. The first step taken by the Porte in this direction was the establishment of the office of Dervendji-bashi, or Grand Superintendent of Roads, whose duty it was to maintain the communications throughout the country, with a number of officers under him, bearing the title of Derven-aga, or guardian of the passes. The creation of such an authority was only a reasonable step, as the means of transit ought always to be at the disposal of the central government; but the employment of Turkish soldiery in their service, which naturally followed, had the effect of raising up another armed force, which was likely sooner or later to come into conflict with the Armatoles. Some time, however, elapsed before any serious differences arose between them; and the two systems might long have continued to exist side by side, had the office of Dervendji-bashi been conferred only on persons of Ottoman extraction. But it has always been the policy of the Turkish power to employ the mutual animosity of its subject nationalities as a means of strengthening its own position; and in accordance with this, about the middle of the last century the plan was adopted of putting that office into the hands of Albanian chieftains,—the leaders of a people whose profession was war, and the bitter and hereditary enemies of the Greek race. From that time a fierce struggle ensued between the Greek militia and the Albanian soldiery, who unrelentingly pursued their object of persecuting and suppressing them. The Armatoles, however, succeeded on the whole in maintaining their position, until the time of the fifth in order of the Pashas of that race on whom the office was conferred: this was the well-known Ali Pasha of Yanina.

Having thus described the character and position of the Armatoles, we may now turn to the Clefts and their relation to them. We have already seen that when the rest of the mountaineers submitted to the Turkish power, a certain number preferred to remain as outlaws in the mountain strongholds, and to maintain their independence in a wild life surrounded by privations. Accordingly the name of Cleft, or robber, which they received signified something very different from ordinary brigands, as it was only the conquerors and oppressors of their brethren who were the objects of their animosity and their marauding expeditions; and with the subject race they usually maintained a good understanding, and were regarded as the assertors of their liberties in time of need. But in the course of time this name obtained a wider signification in the following manner. As long as the rights of the Armatoles were respected by the Turks, they served, as we have said, to maintain order throughout the country: but as soon as they were attacked and forced to stand on the defensive, their character was at once changed, and they assumed the attitude of hostility which they had originally taken up against the conquerors. In this way they for the time became Clefts, and both received this title from others and acknowledged it themselves. When peaceful relations were re-established, they resumed their service as Armatoles; but at last the change from one character to the other was made so suddenly and so rapidly, that the two names came to be regarded as interchangeable, and sometimes those who were living in defiance of the Government were called

Armatoles, and sometimes those who were peacefully engaged in maintaining order were called Clefts.

The most famous among all the head-quarters of these bands was Mount Olympus. In that neighbourhood they continued to exist as late as 1830, when Mr. Urquhart visited the country; and the account which he has given of the captain of a band of Armatoles presents a lively picture of the patriarchal authority of those officers, and their summary mode of administering justice. In the ballads the Clefts of Olympus are constantly mentioned. From the heights of that mountain they are described as watching the Albanians who are in pursuit of them below, and its deep gorges afforded excellent hiding-places, or *lemeria*, as they were called (λημέρια, from ὕλη ἡμέρα), which formed the *rendezvous* of the band during the day; while at night, when there was no need of concealment, they either slept in the open air, or sallied forth on some predatory excursion. The following ballad,—which, though rude and abrupt in its composition, is the most popular, perhaps, of all the Romaic songs, and is well known all over Greece and Turkey,—gives an idea of the estimation in which it was held on this account, while the neighbouring Ossa or Kissavo was regarded with contempt, as being Turkish ground. The strange dialogue with which it concludes is introduced, in order to celebrate the valour of the Clefts of Olympus :—

"OLYMPUS.

"A strife divides the mountains twain, Kissavo and Olympus;
One prides himself upon his swords, the other on his muskets:
Then old Olympus turns himself, to Kissavo he turns him:
'Strive not with me, thou Kissavo, trampled by Turkish footsteps;
I am Olympus, famed of old, renowned throughout the nations,
For two-and-forty peaks have I, and two-and-sixty fountains;

A banner floats on every peak, Clefts swarm in all the bushes,
And on my highest mountain-top an eagle has alighted,
That holds within his talons' gripe the head of some brave hero:'

[*The eagle speaks.*]

'Tell me, thou head, what hast thou done, to be thus hardly treated?
What chance has thus ordained for thee to fall into my clutches?'

[*The head answers.*]

'Consume my might of youth, thou bird; batten upon my prowess;
An ell in length thy wings will grow, a span will grow thy talons.
For twelve years long I lived a Cleft on Chasia [15] and Olympus;
At Luros [16] and Xeromeros [17] I served as Armatolos:
Sixty Agás this hand has slain, and burnt their farms and homesteads;
And those that on the ground I left, Osmanlis and Albanians,
Many, in sooth, they are, thou bird, I cannot count their number.
And now at length my turn has come that I should fall in battle.' " [18]

As a general rule, it was only the Turks who suffered from the attacks of the Clefts. They were considered

---

[15] A chain of mountains north of Tricala, forming part of the Cambunian range.

[16] A river in South Epirus.   [17] The western part of Acarnania.

[18] Passow, 'Popularia Carmina Græciæ recentioris.' No. 131. I subjoin the original as a specimen of these ballads:—

"'Ο Ὄλυμπος κι' ὁ Κίσσαβος τὰ δυὸ βουνὰ μαλώνουν,
Τὄνα παινιέται στὰ σπαθιὰ καὶ ἄλλο στὰ τουφέκια·
Γυρίζ' ὁ γέρος Ὄλυμπος καὶ λέγει τοῦ Κισσάβου·
Μή με μαλώνεις Κίσσαβε μπρὲ Τουρκοπατημένη,
Ἐγώμ' ὁ γέρος Ὄλυμπος στὸν κόσμο ξακουσμένος.
Ἔχω σαράντα δυὸ κορφαῖς κ' ἐξῆντα δυὸ βρυσούλαις.
Κάθε κορφὴ καὶ φλάμπουρο, κάθε κλαδὶ καὶ κλέφτης,
Καὶ στὴν ψηλή μου τὴν κορφὴν αἴτὸς εἶν' καθισμένος,
Ὁποῦ κρατεῖ στὰ νύχια του κεφάλ' ἀντρειωμένου·
Κεφάλι τεῖναι πώκαμες κ' εἴσαι κριματισμένο ;
Πῶς σοῦρτε κ' ἐκατάντησες στὰ νύχια τὰ δικά μου ;
Φάγε πουλί τὰ νιάτα μου, φάγε καὶ τὴν ἀντρειά μου,
Νὰ κάμῃς πῆχυ τὸ φτερὸ καὶ πιθαμὴ τὸ νύχι·
Στὰ Χάσια καὶ στὸν Ὄλυμπο δώδεκα χρόνους κλέφτης,
Στὸ Λοῦρο, στὸ Ξηρόμερο ἀρματωλὸς ἐστάθην·
Ἐξῆντ' ἀγάδαις σκότωσα κ' ἔκαψα τὰ χωριά τους·
Κι' ὅσους στὸν τόπον ἄφησα καὶ Τούρκους κι' Ἀρβανίταις.
Εἶναι πολλοί, πουλάκι μου, καὶ μετριμοὺς δὲν ἔχουν·
Τώρ' ἐρθ' ἀράδα μου κ' ἐμὲ στὸν πόλεμο νὰ πέσω.''

their rightful prey, and were constantly the object of their incursions, sometimes their villages and farms being burnt, while at others the rich owners were themselves carried off to the *lemeri*, and detained there until a sufficient ransom had been paid. But when the robber bands were reduced to extremities, they used also to plunder the wealthier of their own countrymen who dwelt in the plains, especially the bishops, whom they regarded as agents of the Turks. Thus it is that in some of the ballads we hear of persons called Charamides, that is, brigands and murderers, who inspired the greatest terror into the merchants and owners of property. The story of one of the most pathetic songs turns on a recognition by one of those who bore this name of his brother, a young trader, whom he had mortally wounded. But these were quite exceptional cases, and the lower classes, or, in other words, all but a very few of the Greek population, were always on good terms with them, and ready to lend them a helping hand. The shepherds especially are spoken of as the close allies of the Clefts. In this respect they are to be wholly distinguished from the ordinary brigands of the present day, who are the terror and aversion of the people at large. And notwithstanding the acts of violence which at times they did not scruple to commit, they seem to have been usually humane and generous, especially in their treatment of women, who were always safe in their hands, whatever their race or creed. Nor did they ever indulge in those barbarous cruelties which the Turkish Pashas were wont to practise towards any of their number who were taken prisoners. To all this must be added, in explanation of their great popularity, the fame of their extraordinary prowess, skill, and endurance. Their accuracy as marksmen, whether in ordinary mountain warfare by day, or in noting the position of an enemy by the flash of his gun by night; their wonderful strength and agility, arising from their wild life in the open air, and the gymnastic exercises to which their leisure hours were devoted; their power of supporting hunger, thirst, and want of sleep during successive days and nights of combat; and the courage with which they faced death and endured the horrible tortures to which they were exposed as prisoners,—all tended to make them objects of enthusiastic admiration to their countrymen. When the winter arrived, and the mountains were no longer habitable on account of the cold, the Cleft concealed his arms in the *kmeri*, and descended to the lowlands, where he took refuge either in some safe hiding-place on the mainland, or, as was more commonly the case, in the Ionian Islands, which at that time were subject to Venice. But wherever he appeared, he attracted attention and curiosity as one who had defied the common enemy, and slain a multitude of foes. His praises were celebrated in numerous ballads, which circulated from mouth to mouth among the people. Even the children in their games used to play at Turks and Clefts, dividing themselves into two bands, to represent the two opposing parties. The fine little song entitled 'The Cleft's Arms,' of which a translation is here given, may serve to illustrate the feeling that was entertained towards them:—

"The hero's arms! the hero's arms! they never should be sold;
But borne within the church's walls, and blest with solemn tones;
Then hung on high in the ruined tower, the spider's tangled hold,
That the rust may consume the hero's arms, and the earth the hero's bones."[19]

To return now to the history of the Armatoles. Until the time of Ali Pasha this militia had maintained its

---

[19] Passow, 'Pop. Carm.', No. 135.

original organization, and though now and then they were forced to exchange their ordinary life for that of Clefts in the mountains, yet usually it was in the character of Armatoles that they defended themselves and their districts against the encroachments or attacks of their enemies. But towards the end of the century, when Ali was nominated Dervendji-bashi, the aspect of affairs was changed. That chieftain set to work in good earnest to reduce the whole country, and destroy what remained to it of local self-government, employing for that purpose all the forces at his disposal, and all the artifices of which he was master. The immediate consequence of this policy was that all the bodies of Armatoles were for the time broken up, and their members forced to live as outlaws. For a time it seemed as if the Pasha would be as successful in reducing the Clefts, as he had been instrumental in forcing the Armatoles to betake themselves to that kind of life. His Albanian soldiery, being themselves active mountaineers, and acquainted with all the devices of guerilla warfare, were well suited to this kind of service, and, from their great superiority in numbers, succeeded in driving their adversaries from one position to another, and reducing them to great straits. At the same time every kind of stratagem was set on foot to obtain the desired end. Threats, offers of pardon, promises to the chiefs of restoration to their offices, were constantly made with the view of obtaining their submission, and in not a few cases succeeded in doing so. But no sooner was the Pasha's object effected, than every agreement was violated; and assassination was constantly resorted to, in order to remove those who had listened to his proposals. By these means even the Clefts of Olympus were for a time reduced. A ballad still exists which contains the summons which was put forth to them, and which almost all of them obeyed. It runs thus:—

"During this spring-time, during this summer,
 They send to us white papers with black letters;
 'All ye Clefts, who are in the lofty mountains,
 Descend every one of you from Olympus,
 And submit all of you to Ali Pasha.'
 Two Palikars alone did not submit.
 They took their guns, they took their shining swords,
 They ascend to the mountains, and betake themselves to the life of Clefts."[20]

But the moment in which Ali might have thought his work was accomplished, was in reality the beginning of a new era for his opponents. His treachery, cruelty, and extortion had convinced the Greeks generally that submission was only the beginning of a train of worse evils, and that their best hope lay in open rebellion. The consequence was that those Clefts who remained were soon joined by numbers of outlaws from all parts of the country, rendered desperate by ill-treatment, or by the desire of vengeance for those of their relations who had fallen victims to the tyrant. The following song, which belongs to this period, well expresses the prevailing feeling, and at the same time illustrates the spirit of self-assertion, closely resembling that of the Homeric heroes, which characterised these warriors:—

"What, if the Turks the passes hold? What, if the Albanians seized them?
 Yet Stergios is still alive, and he defies the Pashas.
 Long as the snow is on the heights, and flowers in the meadows,
 And fountains in the rocky glens, we'll ne'er be Turkish bondsmen.
 Away! away to our mountain home, the lair that wolves inhabit,
 To the craggy peaks, the hollow caves, the untrodden precipices.

---

[20] Fauriel, 'Chants Populaires,' i. p. 124.

> The dwellers in the towns are slaves: *they* are the Turkish bondsmen;
> *Our* home is in the wilderness, among the savage gorges.
> Live with the Turks! 'twere better far to live with wolves and eagles." [21]

It was at this period that the movement throughout the country began to assume a thoroughly national character. The crowds of refugees who now escaped to the mountains formed a population to themselves, with new interests and new ideas; and among them the conception first arose that Greece might shake off the yoke of the oppressor, and her sons become once more a free people. It was not long before Ali himself recognised what a power he had raised up in opposition to him, and with great sagacity he endeavoured to use it for his own purposes. In 1805 he called together once more the captains of the Clefts, who were now in a position to treat on equal terms, and submitted to them proposals of peace and alliance. On that occasion his lieutenant, Yusuf, the most formidable enemy of the insurgents, astonished at the number of their forces, is said to have inquired of one of their captains, called Athanasius, how it came to pass that after so many years of fighting, and their continual losses, their bands were more numerous than ever. The answer was this:—" Do you see those five young men standing there in front of my Palicars on the right? Two of those are the brothers, two others the cousins, and the fifth a friend of one of my soldiers whom you killed in an engagement. All five of them presented themselves at once to avenge the death of their relation and friend. A few years more of persecution and war, and all Greece will be in our ranks!" This prophecy was fulfilled. The agreement ratified at that time was soon broken,

[21] Passow, No. 54.

owing to the faithlessness of Ali; but when, at a later period, he saw the forces of the Sultan arrayed against him, and his former sources of power failing him, as a last resource he called to his aid these his inveterate enemies, and presented himself to them as their leader on the road to independence. The Pasha himself was overthrown, but not so the cause to which he had unwittingly given an impulse. The bands which had been first brought into existence by his persecutions, and afterwards roused to united action by his call, became the nucleus of the armies which fought in the War of Independence, and won at last the freedom of Greece. At the conclusion of that war the Armatoles, and together with them the Clefts, properly so called, disappeared from the scene. The Armatoles of Mount Olympus, so the hegumen of St. Dionysius informed me, were disbanded about the year 1833, and since that time Albanian guards have formed the police of the mountain.

CHAPTER XX.

TEMPE AND OSSA.

Descent to the Peneius—History of the Koniarates—Site of Gonnus—Ambelakia—Its former Commercial Prosperity—Causes of its Downfal—Vale of Tempe—"The Beauty's Tower"—Ancient Descriptions of the Pass—Ballad of the Salamvria—Monastery of St. Demetrius—Unfavourable Reception—Handsome Central Church—Commencement of Decay—Seacoast of Ossa—Plain and Town of Aghia.

THE descent continued for three hours by a very steep and rough path before we reached the valley—a sufficient proof of the elevated position of the lake, and the impossibility even for a consul Marcius of penetrating into the interior of Olympus from this side. On the way we passed a woman whom we had seen in the cottage we occupied at Nezero, where she had been constantly moaning with pain from the bite of a serpent. She was now on her way to Baba at the entrance of Tempe, where she hoped to be cured by the mudir, who had some skill in medicine. We were told that such bites were not uncommon. At last we found ourselves in the region of plane-trees and agnus castus, in the midst of which a cypress, the first we had seen since approaching Olympus, proclaimed the neighbourhood of the Turkish village of Dereli. Olympus is still a Christian mountain, for the Turks have never established any settlements there, preferring to occupy the more productive plains, and avoiding a continual struggle for an unremunerative possession. Dereli, which lies in the plain, is inhabited by Koniarates, and was one of the original positions occupied by that tribe. Their history is as follows. Shortly before the taking of Constantinople by the Turks the Greek inhabitants of Larissa were reduced to so weak a condition, and had so little hope of assistance from their Government, that they were obliged to submit to the rule of a Bulgarian Prince. Wearied, however, of this yoke, they called in to their aid an Ottoman chieftain called Turkhan Bey, who, after driving out the intruder, established himself in the country. The forces at his disposal being too small to hold it with safety, he sent to Konieh (Iconium) in Asia Minor to invite any of the population who were willing to leave their homes and colonize these fertile plains. Some five or six thousand families responded to his call, and were planted by him in twelve villages under the south-west angle of Olympus, so as to form a barrier to keep in check the Christian mountaineers.[1] Since that time they have greatly increased in numbers, so that they compose the greater part of the Mahometan population of the east of Thessaly. They have a good reputation for industry, but are also said to be extremely fanatical, and we ourselves found them surlier and more insolent than any other Turks, forming a marked contrast to their countrymen of Asia Minor generally, who are quite the most favourable specimens of their race.

The western side of the vale of Dereli, which extends as far as the Peneius, is bounded by a conspicuous spur called Kondovuni; towards the east only one eminence, and that of no great height, intervenes in the direction of Tempe, forming the extremity of one of the buttresses of Olympus. This is called Lykostomo, or "the Wolf's

---

[1] *See* Urquhart, 'Spirit of the East,' i. 335, foll.

mouth," a name which has descended from Byzantine times, when the place was occupied by a town called Lycostomium, but originally, no doubt, applied to the pass itself, which is said by Leake sometimes to bear it still.[2] As it commands the entrance of Tempe from the side of Thessaly it is a position of great importance, and there can be no doubt that it was the site of the Hellenic city of Gonnus. It is surmounted by the ruins of a Byzantine castle, in which I saw no traces of Hellenic walls; but as, unfortunately, I passed it by somewhat inadvertently, I had rather trust M. Heuzey's account, who says that he discovered ancient work, though for the most part composed of small pieces of stone.[3] At this point, according to Herodotus, the route by which Xerxes entered Thessaly to the west of Olympus, debouched into the plain:[4] but it had always seemed to me a difficulty that Gonnus lies so far eastward of that route, and this is only increased by being on the spot, where the ridge of Kondovuni is seen to intervene, lying directly in the way. The regular exit is by a pass further to the west, now called the pass of Meluna; and the only explanation that I know of Herodotus's statement is that suggested by M. Heuzey, viz., that Xerxes had to turn aside in order to occupy Tempe, and that this is loosely described by the historian as entering the plain near Gonnus.[5]

Opposite the Turkish village of Baba we crossed the Salamvria (Peneius) by a ferry: its stream is narrow in this part, not more than 100 feet in breadth, and it appeared to us a small river, after having been so long accustomed to the Vardar. Leaving the main road from Salonica to Larissa, which passes through the

[2] 'Northern Greece,' iii. 384.   [3] 'Le Mont Olympe,' p. 12.
[4] Herod., vii. 128, 173.   [5] P. 17.

village, we began immediately to ascend the flank of Ossa in the direction of Ambelakia, which was the next object of our journey. This place, which was once a manufacturing town of considerable importance, is situated near the head of a steep valley, which has been well compared in shape to an ancient theatre. Its position is charming, for it is upwards of 800 feet above the river, with a northern aspect, and embowered in beautiful trees, plane, fig, and chestnut, while the heights which rise behind it are clothed with oaks. The vineyards, from which it takes its name, and which produce an excellent wine, are seen at every turn, both at the sides of the village, where their bright verdure presents a delightfully refreshing scene, and down the long slopes and terraces below. The entrance of the classic vale is also seen, and beyond it, somewhat further to the east than Ambelakia, the village of Rapsani, lying on the side of the heights of Olympus. The aspect of Ambelakia from without is imposing, and it numbers 300 houses, and has two schools; but many of the handsomest dwellings are in ruins or in a state of decay, and consequently an air of melancholy seems to pervade the whole place. Still the appearance of the people is superior to what is usually seen in Turkey, and the women are decidedly good-looking. But its commerce is wholly a thing of the past, as is also the case with that of the other Christian communities of Ossa and Pelion. The particular branch of trade for which Ambelakia was famous was that of dyeing thread. At the present time there is one dyeing house for the blue dye, which is used for the thread employed in the looms of the country; but the red dye, derived from madder (*rizari*), with which the thread was formerly dyed for exportation to Germany, is now unknown.

At the beginning of this century, when this village was at the height of its prosperity, it was visited by several travellers, and among them by Beaujour, Clarke, and Leake, who have left us an account of its state at that time.[6] The last-named traveller, with his accustomed admirable accuracy, has given details respecting the process of dyeing, and statistics as to the exports and the system on which the trade was managed. From these notices we learn that the madder was imported through Smyrna from Asia Minor, where it grows wild on the mountains, in which state it produces a finer colour than when cultivated. The thread, on the other hand, was procured from the neighbouring parts of Thessaly, and partly spun by the women and children of the place. Of this, when dyed, as many as 150,000 to 200,000 okes[7] were exported to Germany every year, being carried on the backs of horses overland to Belgrade. The expense of carriage necessarily decreased the ultimate profit, but nevertheless this was at one time considerable, from the economical way in which the business was conducted, when all the inhabitants formed one company, even the lowest taking part in the work and enjoying his share. At the same time their agents abroad were members of their own community, by means of whom the profits of brokerage were secured to the company. Beaujour's highly-coloured account of their system of administration has been often quoted, but as it is remarkably interesting I cannot do better than insert it here.

"Ambelakia, from its activity, resembles rather a town of Holland than a village of Turkey. This village, by its industry, spreads move-

---

[6] Beaujour, 'Tableau du Commerce de la Grèce,' i. p. 272. Clarke, 'Travels,' iv. p. 285. Leake, 'Northern Greece,' iii. p. 385.

[7] The oke is somewhat less than three pounds weight.

ment and life throughout the surrounding country, and gives birth to an immense commerce, which links Germany with Greece by a thousand threads. Its population has trebled in fifteen years, and at the present time (1798) amounts to 4000 souls, all of whom live in the dyeing-houses, like a swarm of bees in a hive. In this village both the vices and the cares which idleness begets are unknown. The hearts of the Ambelakiotes are pure, and their countenances serene. The slavery which blasts at their feet the plains watered by the Peneius, has not ascended their hill-sides; no Turk is permitted to dwell or sojourn among them, and they are governed, like their forefathers, by their primates and their own magistrates. Twice the fierce Mussulmans of Larissa, jealous of their comfort and prosperity, have attempted to scale their mountains and pillage their houses; twice they were repulsed by hands which at a moment's notice dropped the shuttle to arm themselves with the musket.

"Every hand, even the children, are employed in the factories of Ambelakia; and while the men dye the cotton, the women spin and prepare it. The use of the spinning-wheel is unknown in this district of Greece; only the spindle is employed. The thread is undoubtedly less strong, less round, and less equal; but it is softer, firmer, and more silky. It is less apt to break and more durable, is more easily bleached, and takes the dye better.

"At Ambelakia there are twenty-four manufactories, in which 6138 cwts. of cotton-yarn are dyed yearly. All this finds its way into Germany, and is distributed between Pesth, Vienna, Leipzig, Dresden, Anspach, and Bayreuth. The merchants of Ambelakia have houses in all these towns, where they sell the yarn to the German manufacturers. Originally, all these houses were managed by a number of societies with independent interests; but, as these injured one another by competition, they conceived the idea of combining them all into one. Twenty years ago the plan of a general administration was formed, and one year after it was put into execution. The rules of the new company were drawn up by men of judgment. Each proprietor or head of a factory might contribute proportionately to his means. The lowest shares were fixed at 5000 piastres, and the highest were restricted to 20,000, in order that the rich might not have it in their power to swallow up all the profits. The workmen might unite their savings, and took shares by subscribing among themselves, thus forming small companies incorporated in the great one. Besides their money, these workmen gave also their labour and skill; and their wages, together with the interest of their investments, soon diffused comfort through every family. The returns of the dividend were restricted to 10 per cent., and the surplus was devoted to increasing the original capital, which in two years rose from 600,000 piastres to 1,000,000. Never was any society established on more economical principles, and never were fewer hands employed to carry on a business of the same magnitude.

"For a long time the administration was carried on with the most perfect harmony. All the members vied with one another in contributing to its success. The directors were disinterested, the correspondents zealous, the workmen hardworking and docile.

"From being apportioned equally to all the workmen and factories, the work was performed with carefulness and rapidity, and all the houses flourished. The profits of the company increased every day, until an immense capital was formed."

Several causes combined to bring about the ruin of this flourishing community. In the first place, the increasing scarcity of madder added considerably to the expense of the manufacture. Then came dissensions at home, leading to the disruption of the company; and at last the finishing stroke was given by commercial failures in Germany, accompanied by other unfortunate circumstances. The history of these may be given in the words of Mr. Urquhart. After noticing the absence of any judicial authority among them, by which disputed questions might be decided, he continues:—

"The infraction of an injudicious bye-law gave rise to litigation, by which the community was split into two factions. For several years, at an enormous expense, they went about to Constantinople, Salonica, and Vienna, transporting witnesses and mendicating legal decisions, to reject them when obtained; and the company separated into as many parts as there were associations of workmen in the original firm. At this period the Bank of Vienna, in which their funds were deposited, broke; and, with this misfortune, political events combined to overshadow the fortunes of Ambelakia, when prosperity and even hope were finally extinguished by the commercial revolution produced by the spinning-jennies of England. Turkey now ceased to supply Germany with yarn; she became tributary for this, her staple manufacture, to England. Finally, came the Greek Revolution. This event has reduced, within the same period, to a state of as complete desola-

VOL. II. F

tion the other flourishing townships of Magnesia, Pelion, Ossa, and Olympus." [a]

But the speedy collapse of such a community is not so much a cause of wonder as the fact of it having existed at all. External circumstances can hardly be said to have contributed to its rise or its prosperity, for, with the exception of retirement and security, it derived no advantages from its position—or rather, one might say, every disadvantage, being without a port, far removed from the market where its goods were sold, and out of reach of the stimulus supplied by civilisation and mercantile neighbours. It is an example of what may be effected with very small appliances by an enterprising spirit and untiring industry, combined with hearty co-operation and a compact and economical system.

From Ambelakia we descended on the opposite side of the valley from that by which we had approached it, until we reached a small khan near the river bank, under a spur which here projects from Ossa; at this point the pass of Tempe properly begins. The term "pass," which I have so frequently applied to it, and its modern name, "the Wolf's Mouth," will have prepared the reader for something very different from the gentle pastoral valley which it has often been conceived to be: it is, in fact, a deep chasm—cloven in the rocks, as the fable tells, by the trident of Neptune—between Olympus and Ossa; and it was a work worthy of the earth-shaking god, for without it the Peneius would have had no outlet, and Thessaly would have been, what perhaps it once was, a vast lake. At the same time, though it possesses every element of the sublime, it has also many soft and beautiful features, in the broad winding river,

[a] Urquhart's 'Spirit of the East,' ii. 20, 21.

the luxuriant vegetation, and the glades that at intervals
open out at the foot of the cliffs, which distinguish it
from ordinary passes, and enable us to recognise in it the
Tempe of the poets. The track follows the right bank
of the stream, and just within the entrance, where it
passes over the bare rocks, the marks of chariot-wheels
show evident traces of the ancient road. Beyond this
the valley closes in, and the rocks become higher and
higher at every turn, those towards Olympus being the
steepest, so as completely to bar the passage on that
side of the water, while those which descend from Ossa
are the highest, rising in many places not less than
1500 feet from the valley. Further on still there is seen,
by the river bank below the present road, a wall composed of large blocks, which served to support the
ancient road. Towards the middle of the pass the
precipices on either side approach nearer and become
extremely steep, only just leaving room for the road and
the river; in the neighbourhood of these we heard the
romantic sound of the shepherd's pipe, and, when one of
our party shouted, the echoes answered finely from the
cliffs. Here there are remains of two mediæval castles,
which served for the defence of the pass: one of them,
which stands close above the road, is evidently from its
position on the site of the Hellenic fort mentioned by
Livy, who describes its position as being in the narrowest
part of the valley, towards its middle, where the road
might easily be defended even by ten men.[9] M. Mézières, in his excellent book on 'Pelion and Ossa,'[10]
expresses his opinion that the foundations of the existing
building are Hellenic; in this, however, he is mistaken,

[9] Livy, xliv. 6.
[10] 'Mémoire sur le Pélion et l'Ossa,' p. 115. This little work forms a
very complete guide to the two mountains.

for there is no trace of ancient work in any part of it. The other castle, which surmounts the cliffs at a great height above, is a strikingly picturesque object, the most conspicuous portion of it forming an arch which stands out against the sky. It is called "the Beauty's Tower" (κάστρο τῆς ὡραίας), a name which is found elsewhere in Greece, and seems to be connected with a story embodied in several Greek ballads, of a Christian lady whose fortress was captured by means of a stratagem by a band of Turks in the following manner. They hide themselves in ambush near the entrance gate, while their leader, who is dressed as a monk, approaches, and begs to be admitted to receive charity and to worship in the chapel. After a long parley the door is half opened, on which the infidels rush in and pillage the place. The heroine, who is represented as throwing herself headlong from a window to escape from falling into the hands of her captors, is called "the Beauty of the Tower" (ἡ ὡραία τοῦ κάστρου).[11]

Immediately beyond these castles a grand and wild gorge descends from the heart of Ossa, and both here and in the opposite precipices of Olympus are seen the openings of caverns, which probably were once inhabited by ascetics. The cliffs all through the pass are composed of grey limestone finely tinted with red, and their ledges and hollows are fringed with trees which fix their roots to the rocks. The vegetation is magnificent, and, wherever the slopes are sufficiently gradual, runs far up the mountain sides: it is composed of oak, wild olive, and dwarf ilex, together with a thick undergrowth of agnus castus, palluria and oleander, while the banks of the stream are everywhere shaded by plane-trees of

---

[11] Passow, 'Carmina Popularia,' Nos. 485, *sqq.*

luxuriant growth. In a few places also may be seen the laurel of Apollo, which that divinity was said to have transplanted from hence to Delphi; in memory of this event a sacred deputation was sent every ninth year by the Delphians to offer sacrifices and cull fresh branches from the trees. The water of the Peneius, though not clear, is pleasing to the eye from its pale green colour, and very different from what it appears in the spring, when it is white and turbid, and has caused a feeling of disappointment in the minds of most travellers. It is strange that of the numerous descriptions of the valley by ancient writers that have come down to us, not one can be considered really accurate. Catullus, with his usual felicity, has seized on one salient feature in his "Tempe girt with overhanging woods;"[12] but those who have attempted to depict it more fully, whether in prose or verse, seem either to draw on their imagination or to miss the most characteristic points in the scene. Thus Livy, whose account is probably derived, like his narrative of the consul Marcius' expedition, from the eye-witness of Polybius, dwells exclusively on the steepness of the precipices and the roar of the river,[13] the latter feature being an impossible one, for the Peneius is too broad, and its bed too deep, to admit of much rushing sound being produced by its current: this idea Ovid has elaborated with much poetical exaggeration.[14] Pliny, again, had spoiled a description otherwise satisfactory, by speaking of the mountains as "gently sloping."[15] The truest, and at the same time the most picturesque

---

[12] "Tempe, quæ silvæ cingunt superimpendentes."—'Nupt. Pel. et Thet.,' 286.

[13] "Rupes utrinque ita abscisæ sunt ut despici vix sine vertigine quádam simul oculorum animique possit. Terret et sonitus et altitudo per mediam vallem fluentis Penei amnis."—Livy, xliv. 6.

[14] Ov. 'Met.,' i. 568.   [15] "Leniter convexis jugis."—Plin., iv. 8.

account, though somewhat encumbered by details, is that of Ælian;[16] but even this leaves on the mind an incorrect impression as to the breadth of the river, from his describing the trees on the banks as screening those who sail down the stream from the rays of the sun.

As we proceed, another grand buttress is thrown out from Ossa, and descends with such steepness to the river, that the road is forced to quit the bank and cross its shoulder. At the commencement of the ascent the rocks have been cut away to form the ancient road, and are ribbed and grooved in a manner similar to those at the Demirkapu of the Vardar. Near this point there is a Roman inscription on the rocks, placed there by one of Cæsar's lieutenants; both my companion and myself searched carefully for it, but in vain. It is given by Leake, and runs as follows—*L. Cassius Longinus Pro. Cos. Tempe munivit.* When the highest point is attained the views both up and down the pass are most superb. Looking back, you see a long reach of the tranquil river below you, enclosed on both sides by luxuriant woods, and backed by a succession of towering cliffs; in the opposite direction the mouth of the valley appears, and beyond it a fertile plain covered with trees, while the horizon is bounded by the faint line of the hills of Chalcidice. At the exit of the pass, which is four miles and a half in its entire length, we found a place for our midday halt which completely realised the Tempe of our imagination. At the foot of the rocks, which here approach the river, a copious stream of clear water gushes from numerous sources beneath the shade of gigantic plane-trees, in the midst of beds of spreading fern; to make its pastoral character quite complete a

---

[16] 'Var. Hist.,' iii. 1.

VALE OF TEMPE.

herd of goats was browsing in the neighbourhood. Here we dined and reposed in a state of great enjoyment, after a previous plunge in the classic river.

It is remarkable that, in four of the Romaic ballads, where the Salamvria is mentioned, it is spoken of as a river of the dead, and as having something of the same power of inspiring forgetfulness which was attributed by the ancients to the water of Lethe. Can it be supposed to have inherited in the minds of the people the "infernal" character which once belonged to its tributary the Titaresius? Of that stream Homer says that it refuses to mix its waters with those of the Peneius, and flows over it like oil, because it is a branch of the dreadful Styx;[17] and Pliny even calls it by the name of Orcus.[18] When Colonel Leake visited the waterfall of the Styx in Arcadia, he found some of the superstitions still remaining which Pausanias mentions as attaching to it; so that it is conceivable that here also some such tradition may exist. The following is a translation of the best of the ballads that relate to this subject; like most of the others it is often abrupt in its transitions, but is not without an element of poetry.

> "The mother of two gallant sons, who ply their trade as fishers—
>   Go, bid her wait for them no more, no more come forth to meet them:
> Salamvria's stream came down amain, by sunlight and by moonlight,
> And they are tossing in its depths, like fishes in the ocean.
> 'O clasp me, clasp me, brother dear! one tender kiss at parting!
> It may be we shall meet again, and welcome one another.'
> 'When the dry trunk puts forth new buds, and clothes itself with branches;

---

[17] 'Il.,' ii. 753, *seqq.* Leake remarks that "the apparent reluctance of the water of the Titaresius to join with that of the Peneius arises from the former being clear, and the latter muddy."—*Northern Greece*, iii. 896. Strabo just inverts Homer's statement, and makes the Peneius the clear river, being misled apparently by the epithet ἀργυροδίνῃ.   [18] Pliny, iv. 8.

When the crow shines in plumage white, the snowy dove to rival,
O then may we two meet again, and welcome one another;
Then may we go, and not till then, together to our mother.'"[19]

Just below the point where the river enters the plain there is a ferry, by which it is crossed to go to Salonica; and some distance further down its stream is spanned by a stone bridge of several arches, now ruined and disused. Our route lay close under the seaward slopes of Ossa, through a rich plain covered with fine maize plantations and well-grown trees, which in ancient times belonged to the district of Magnesia.[20] At the distance of three miles from the end of the pass is a place called Laspochori (mud village), where we were told that there are old walls on the hill above, which is a good position for a city. This was most probably the site of Homoloium, which is stated by Strabo to have been under Ossa, close to the exit of the Peneius from Tempe.[21] The plain extends towards the south for seven miles further, where at its extremity there is a little port or *scala* called Chaizi. The name of Fteri has been given to this place in the maps, apparently on the authority of Leake, who did not visit these parts, but obtained information about them; however, I was told, in answer to my repeated enquiries, that no such name was known in the neighbourhood. It signifies "fern," and the mistake may have arisen from the abundance of that plant in the plain.

By the side of a ravine, on the mountain slope above the *scala*, commanding an extensive view over the Ægean, is the ancient monastery of St. Demetrius, of which we had often heard in the course of our wanderings

---

[19] Passow, 'Pop. Carm.,' No. 388. Compare also No. 371, where a river of forgetfulness is spoken of as existing in "the lower world."
[20] Strabo, ix. 5, 22. [21] Ibid.

on Olympus. We ascended to it through a forest of planes and chestnuts, but on our arrival were admitted with some reluctance, as the hegumen, whom to our surprise we found to be the only monk, was not there, and might not return that evening. However, as we were extremely anxious to see the antiquities of the place, we determined to remain for the present, and gave orders that our baggage-horse should be unloaded. On returning from the further end of the building, where I had been examining the exterior of the central church, I heard the sound of loud voices, which proved to proceed from our dragoman and a woman with a sour countenance but rather handsome features, who was pouring forth a torrent of abuse at the top of her voice against our whole party. The cause of this was soon explained. She had heard that the cholera was raging in Constantinople, and was impressed with the idea that we had escaped from that place, and were bringing the infection with us. We protested our innocence, and explained that we had been travelling in the healthy climate of Mount Olympus; but arguments and protestations were of no avail. Now a shrew in a passion is never a very agreeable sight; but the unpleasantness is greatly increased when you know that she may have it in her power to deprive you of your night's lodging, not to say your supper. And this appeared to be the case, for we found that she was sister-in-law to the hegumen, and all-powerful in his absence. At last, when her flow of language was exhausted, she gathered her children round her—as if to save them from contagion—and, with the air of a Medea mounting her dragon-car, ascended the steps that led to the upper story, and then locked and double-locked herself and her family into one of the chambers. The aspect of affairs was certainly very unpromising, and we were half-inclined to leave the place, when, to our great joy, the hegumen arrived and gave us the heartiest welcome. He was a very small man, and almost a hunchback, but far above the ordinary run of caloyers in intelligence and information. Originally he had belonged to the monastery of St. Dionysius on Olympus, and he had been on a visit in the neighbourhood of that place when we were there: so he said he was in hopes we should pass in the direction of his monastery, and should have been much disappointed if he had missed us. Accordingly we were soon installed in a comfortable room, and a sumptuous supper was provided for us. The quick-tempered lady, it need hardly be said, did not reappear. "Ah, well!" the little man philosophically remarked, "she is a woman!"

The founder of the monastery, according to our host, was the Emperor Justinian, but so many ecclesiastical institutions are referred to him by tradition, that we may hesitate to believe it. However, in an inscription built into the wall, in one part of the present structure, it is stated to have been founded in the year A.D. 575 (+ ετους ἐξε εκτιςη, [sic]),[22] which will correspond to the date of his successor, Justin II. The buildings were in all probability at one time much more extensive than those now existing, which form three sides of a quadrangle, the fourth being occupied by a blank wall. None of these are of any antiquity, except the central church and the kitchen, the latter of which is built on the usual type of monastic kitchens, being square, with a large central

---

[22] The notation of this number may excite suspicion, but there is evidence to show that the Arabic system of computation was in use among the Greeks and Romans before the end of the fifth century. At the same time, it does not follow that this inscription is coeval with the foundation of the monastery, though it is certainly ancient.

hearth; and over this in the middle of the roof is the chimney, which here takes the form of a Byzantine cupola. Close to it once stood the refectory, but this is now destroyed. The church, the accompanying plan of which is taken from M. Mézières' work, is a very fine and very ancient structure, though the idea that it dates

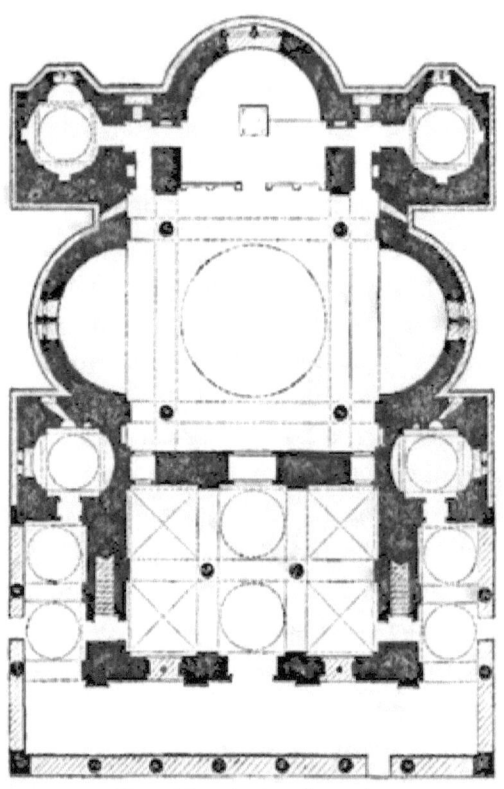

Plan of Church of St. Demetrius.

from the 4th century, which is put forward by that writer, or rather by his companion M. Normand, to whom he refers on subjects of art, is quite extravagant. The history of Byzantine architecture has been so little studied, that it is as yet impossible without historical evidence to fix the exact period of any building in that style; but it is not at all probable that this church is even coeval with the date given above for the foundation of the monastery, for it has no mosaics or other features by which the earliest Byzantine buildings are characterised, nor is there anything to distinguish it from many buildings on Athos confessedly of a later period. The fact that the eastern end is angular on the outside, and not semicircular, is, however, a sign of some antiquity. The only points in which the construction differs from that of ordinary monastic churches are the small chapels at the sides, and the chamber over the *narthex;* the latter of these, however, is found at St. Dionysius on Olympus, and in the church which I have already described at Prisrend. Much of the stonework,—as, for instance, the capitals, the small pillars dividing the windows, and the cornices of the doors,—is elaborately carved in incised work. The interior is grand and lofty, and the effect of this is no doubt increased by its being now almost denuded of church furniture. Of the four pillars that support the central cupola, three are of granite, and one of white marble: these have evidently been taken from an earlier building, for their capitals and bases do not correspond to them in size, or in any other respect; while the bases of those in the *narthex* are formed of capitals turned upside down, and one of those in the *proaulion* or outer porch has a Corinthian capital much too large for it. All this seems to imply that the monastery occupies the site of an ancient building, the materials of which were employed in its construction; and this is confirmed by the pieces of columns and cornices which lie embedded in the pavement of the court, or are built into the modern walls, and look too large to have originally belonged to a monastic edifice. But if Justinian really was its founder, it is quite possible that he brought the spoils of heathen temples here, as he did to St. Sophia's at Constantinople; nor is there any reason for supposing that a city occupied this position, which is an open mountain side without any natural strength.

The frescoes with which the whole of the interior is covered have a fine effect, and those in the cupolas are unusually rich, from the amount of gold ground which has been introduced. None of them, however, are of any antiquity, except perhaps those in the upper part of the east end of the church. It is likely enough that they have been painted over an earlier series of frescoes, but the character of them has been wholly lost, and the delineation of the faces is in a completely modern style. If, as M. Heuzey suggests, in speaking of the ornamentation of some of the churches on Olympus, there was a period of *renaissance* in art throughout these parts, corresponding to the time of the prosperity of Ambelakia and the neighbouring manufacturing towns, they may date from that time. The *iconostase* is very richly carved and gilt, though almost all the icons are now removed; in the floor, too, which in many places has sunk, like that of St. Mark's at Venice, there are pieces of porphyry and traces of *opus Alexandrinum*. But the days of this fine building are numbered. A wide crack extends all down the northern semi-cupola, and the stilting above one of the pillars on that side is wholly out of the perpendicular; and in the course of last year a great piece of the roof of the eastern semi-cupola above the altar fell down, bringing with it the finest of all the frescoes, and was smashed in pieces on the floor, where its ruins still lie, interspersed with gilding and brilliant colours—a sad sight! The hegumen talked of collecting money to repair it; but this can never be.

Now that decay has begun its work it must go on year by year, until the whole falls to pieces: perhaps we are the last travellers who will have seen it in its beauty. The funds on the spot are small indeed, the annual revenue of the monastery having dwindled, owing to a concurrence of adverse circumstances, to the small sum of 50*l.* This accounts for its being occupied by one monk only.

Our route from this place lay towards the south, under the flanks of Ossa, in the midst of exquisitely beautiful scenery, by an unfrequented track, which in places was hard to find, sometimes lying among the trees that fringe the shore, and at others high up on the mountain sides, commanding a wide expanse of the blue waters. At first we descended until we joined the path leading to Karitza, and then ascending again looked back over the monastery, together with the successive ranges of the broad Olympus, the mouths of the Peneius, and the Thermaic gulf. After threading our way through forest land for an hour and forty minutes we reached Karitza, a good-looking place, which is situated inland, and high above the sea; its inhabitants, like those of most of the villages on the seaside of Ossa and Pelion, are nearly all sailors, so that at this period of the year hardly any but women and children are to be seen there. A man, from whom I made enquiries, informed me that on a hill to the south of the place, some way above the road, there were remains of old walls: in all probability these mark the site of Eurymenæ, which seems to have been the most important of the few towns situated in this rugged district; we should have liked to explore them, but in this climate during the month of August acropolis-hunting in the middle of the day is almost an impossibility. All about this part of Ossa, and also on the sides of Olympus, the cornel tree grows abundantly, and its acid fruit is greatly relished by the natives. At last the path leaves the coast, and strikes inland through narrow wooded glens towards the central ridge, which descends from the summit of Ossa towards the foot of Pelion. As you approach it, the rich vegetation ceases, though the scenery is still pretty, the sea being frequently visible at the same time through a number of separate valleys. Below the ridge on its western side lies the village of Thanatu, on descending from which you overlook the plain of Aghia, green with mulberry grounds and maize plantations, which forms, as it were, a bay, running in from the great Thessalian plain, and enclosed between the counterforts of the two great mountains. Of these the most conspicuous are the northern buttresses of Pelion, which almost form a range by themselves, and bear the distinctive name of Mavro-vuni. The town of Aghia, which stands at the head of the plain in a smaller basin of its own, is composed of 400 houses, entirely inhabited by Greeks. In former times it was devoted to the manufacture of two kinds of stuffs, viz., *fitilia*, which was composed of silk only, and *aladja*, of mixed silk and cotton; but though this no longer exists, the articles having been driven out of the market by the competition of European productions, yet the place continues to have a thriving look, for silk is cultivated and exported in large quantities, together with a considerable amount of wine and oil.

CHAPTER XXI.

THE VRYKOLAKA, OR EASTERN VAMPIRE.

Apparitions of Spirits — Terror inspired by the Superstition — Origin of the name Vrykolaka — Connection with Lycanthropy — Malignant Vampires — Frequently produced by Excommunication — The Body undecomposed — Effect of Absolution — Stories relating to it — Modes of Exorcism — Supposed Offspring of Vrykolakas — Burning the Body — Tournefort's Narrative — Innoxious Vampires — Ballad of "The Spectre."

DURING the night which we spent at Aghia the population were disturbed by apparitions of spirits, which they described as gliding about with large lanterns in their hands. These are called *vrykolaka* by the Greeks and *vurkolak* by the Turks, for both Christians and Mahometans believe in them; the name, however, is written and pronounced in a great variety of ways. It was curious to meet with them in this manner as soon as we descended into the plains of Thessaly, the ancient land of witches; but the belief in these appearances is widely spread, not only throughout Thessaly and Epirus, but also among the islands of the Ægean and over a great part of Turkey. The idea concerning them is, that some persons come to life again after death, sleep in their tombs with their eyes open, and wander abroad by night, especially when the moon is shining brightly. In reference to this, there is a form of banter common among the Greeks, in which a man says to his friend—

> "I wish you may never die:
> Or, if you do, that you may become a *vrykolaka*,
> In order that you may enjoy the world twice over."[1]

The more malignant form of these nearly corresponds to the *vampire*, the belief in which was at one time so widely spread throughout Europe, and which was supposed to keep itself in life by sucking the blood of men. The ghouls of the 'Arabian Nights' represent a similar idea. The terror caused by this superstition is intense. Lord Byron mentions in one of the notes to 'The Giaour' that he saw a whole family terrified by the scream of a child, which they imagined must proceed from such a visitation. And Mr. Newton tells us that there is no readier or more effectual way of getting rid of an importunate or tiresome Mytileniote than to say to him, "May the vampire take you:" he immediately crosses himself, and withdraws.[2] Moreover, as the Turks are especially afraid of Christian vampires, and *vice versâ*, it sometimes becomes a vent by which religious animosity can show itself. In Samos, some time ago the Turks were haunted by the spectre of a Greek who had been lately buried, so that ultimately they persuaded the authorities to have his body dug up and burnt; after which the Christians complained that they were not only persecuted while alive, but were not allowed to rest when dead.

As the details of this belief in the East are very curious, and take a variety of forms, I will here introduce a short notice of what may be gathered from various sources about it. Besides the name already given,

[1] "πότε νὰ μὴν ποθάνῃς,
κἂν ποθάνῃς, νὰ βρυκολάσῃς,
διὰ νὰ χαρῇς τὸν κόσμον δύο φοραίς."

[2] 'Discoveries in the Levant,' i. p. 212.

which appears also in Servia as *vukodlak*, and in Albania and elsewhere as *vurvulak*, there are other titles of quite a different origin applied to such apparitions. Thus by the Wallachians they are called *murony*, and in Crete and Rhodes they bear the name of *katakhanás* (καταχανάς), a word which appears originally to have signified "a destroyer."[3] Of the name *vrykolaka* and its cognate forms various derivations have been given at different times. Leo Allatius[4] amusingly compounds *vurkolakas* of βοῦρκα, "foul mad," and λάκκος, "a ditch," as if it represented the scandalous character of such persons as became vampires after death. With greater semblance of probability the modern Greeks, and others who have followed them, derive it from the old Greek word for a hobgoblin, μορμολύκη. Others have thought that the Servian *vukodlak* is the original; and its meaning in that language, viz., "wolf hairy," suggesting as it does a direct connection with the kindred superstition of lycanthropy, seems at first sight to confirm this view. The connection with lycanthropy, as we shall presently see, is distinctly traceable, and there is an etymological connection between the words in these two languages; but the modern Greek form of the word is in reality earlier than the Servian, for the word for a wolf in Sanskrit is *vrka*, and the *r* in that word, which is preserved in *vrykolaka*, is lost in *vukodlak*. The Bohemian *vlkolak* shows us the original form in another Slavonic language, with only a change of liquids. It is highly probable that the Greek name came from a Slavonic source, but the existence of the other name, *katakhanas*, shows that the idea existed among the Greeks independently.

Mr. Baring Gould, in his 'Book of Were-wolves,' has

[3] Newton, i. p. 212. Pashley, 'Travels in Crete,' ii. p. 208, and *note*.
[4] 'De quorundam Græcorum Opinationibus,' p. 142.

spoken of the *Vrykolaka* as if it was identical with the were-wolf,[5] and says that those who are believed to be lycanthropists during life become vampires after death. This, however, is, I think a mistake. In the great majority of instances the were-wolf superstition is wholly independent of this belief; so much so, that one writer, who has carefully collected the authorities on the subject, expresses his opinion that the nature of the were-wolf is no longer to be recognised in the modern Greek *Vrykolaka*.[6] Amongst the Wallachians, however, there is a kind of *murony* that corresponds to the belief in kynanthropy, which is one of the forms of the same superstition. This is described as "a real living man, who has the peculiarity of roaming by night as a dog over heaths, pastures, and even villages, killing with his touch horses, cows, sheep, swine, goats, and other animals in his passage, and appropriating to himself their vital forces, by means of which he has the appearance of being in continual health and vigour."[7] The name of this being, the *priccolitsch*, is evidently another form of *vrykolaka*; from which it is probable that the modern Greek belief was once connected with the same notion, more especially as the idea of lycanthropy was well established among the Greeks in classical times. Indeed, if we may believe M. Cyprien Robert,[8] this same belief is also found as a form of vampirism in Thessaly and Epirus; but his authority is hardly sufficiently trustworthy to be received on such a subject. Another proof of the connection of the two ideas is found in the notion, that one of the causes which convert men into vampires after death is

---

[5] P. 114.
[6] Wachsmuth, 'Das alte Griechenland im neuen,' p. 117.
[7] Schott, 'Walachische Märchen,' p. 298.
[8] 'Les Slaves de Turquie,' i. p. 69.

the eating the flesh of a lamb that has been killed by a wolf.[9] With this may perhaps be compared the belief which exists in some parts of Albania, that this effect is produced on any corpse over which a cat or wild beast has sprung.[10] Without entering further on the question of lycanthropy, we may notice how easy the transition is from the one superstition to the other; for at a very early period in the history of the Indo-European race the wolf, partly as being the great enemy of shepherds, and partly, no doubt, from its sinister appearance and habits, came to be regarded as a representative of the evil powers.[11] Hence the Germans and Slaves have always attributed to the wolf a demon nature; and M. Wachsmuth tells us that he was informed (though I cannot say that this is confirmed by my own observation) that the modern Greeks are in such fear of this animal that they shrink from even pronouncing its name.[12]

I have already spoken of the belief in two forms of *vrykolaka* as existing in Turkey, one more and one less malignant. The blood-sucking vampire is probably the only representative of this Eastern superstition known in England, as it has been popularized by Byron in the lines of 'The Giaour'—

> "But first, on earth, as vampire sent,
> Thy corse shall from its tomb be rent:
> Then ghastly haunt thy native place,
> And suck the blood of all thy race."

---

[9] Wachsmuth, p. 117.
[10] Hahn, 'Albanesische Studien,' i. p. 163. In parts of England it is still considered a very bad omen for a cat or dog to pass over a corpse or coffin; in such cases the animal is immediately killed. See Henderson's 'Folklore of the Northern Counties,' p. 43.
[11] See Pictet, 'Origines Indo-Européennes,' ii. p. 639.
[12] On the true lycanthropy—that is, the very rare form of monomania that consists in a person believing himself to be changed into a wolf, the reader

But in reality the less noxious kind is quite as often found. Speaking of the Albanian *curculak* Von Hahn tells us that none of the apparitions in that country are thought to suck blood or feed on corpses. And in a large number of the stories on the subject which come from Greek sources, they are described as playing all kinds of mischievous tricks, and frightening the people in a variety of ways, but not as causing any further injury. The more malignant ones, according to Pashley, feed especially on the liver of their victims; and this idea is supposed to account for an exclamation which a traveller heard from a Cretan mother: "I will sooner eat the liver of my child."[13] The *murony* of the Wallachians not only sucks blood, but has the power of assuming a variety of shapes, as, for instance, that of a cat, frog, flea, or spider; in consequence of which the ordinary evidence of death caused by the attack of a vampire, namely the mark of a bite on the back of the neck, is not considered indispensible, and the fear of sudden death is indefinitely increased, as it is at once attributed to this cause, and one who has been destroyed by a vampire becomes himself a vampire.[14] In the north of Albania a similar idea exists relatively to the transformation of the wandering spirit, and this is regarded as an additional punishment for previous crimes.[15] Some of the Greeks again believe that the spectre which appears is not really the soul of the deceased, but an evil spirit which enters his body. Thus Leo Allatius, in describing the superstition, says— "The corpse is entered by a demon, which is the source

is referred to Dr. Pusey's 'Daniel the Prophet,' p. 425, foll., where all that is known on the subject has been brought together with the writer's usual learning. [13] Pashley's 'Crete,' ii. p. 200.

[14] Schott, 'Walachische Märchen,' p. 297.

[15] Hecquard, 'La Haute Albanie,' p. 342.

of ruin to unhappy men. For frequently, emerging from the tomb in the form of that body, and roaming about the city and other inhabited places, especially by night, it betakes itself to any house it fancies, and after knocking at the door addresses one of its inmates in a loud tone. If the person answers he is done for; two days after that he dies. If he does not answer, he is safe. In consequence of this, all the people in that island (Chios), if any one calls to them by night, never reply the first time; for if a second call is given then they know that it does not proceed from the *vrykolaka*, but from some one else."[16]

The principal causes which change persons into *vrykolakas* after death are excommunication, heinous sins, the curse of parents, and tampering with magic arts.[17] The first of these is the most common and most important, and dates from very early times. It would seem as if the priests must have made use of the superstition to work on the fears of the people; nor did they confine their power to the state after death, for they have persuaded their flocks to so great an extent of their influence over their health during their lifetime, that in many places, if a man has a headache, fever, or rheumatism, he at once attributes it to the excommunication of a priest. In some cases this belief has reacted injuriously on the clergy themselves; for in one instance a priest was killed in revenge for the death of a man whose illness was gratuitously attributed to his supposed excommunication; and in another, a bishop, who had been robbed by a number of mountaineers, was afterwards pursued and shot by them, lest he should excommunicate them as soon as he reached a place of safety.[18] In all cases,

---

[16] 'De quorundam Græcorum Opinationibus,' p. 142.
[17] Wachsmuth, p. 116.   [18] Pashley, pp. 222, 223.

where a person dies excommunicate, his body refuses to decompose in the grave. Leo Allatius describes a corpse which he himself saw in this condition, and he seems to imply that the Greeks connected this circumstance with the text—" Whatsoever thou shalt bind on earth, shall be bound in heaven," the soul being excluded from all hope of participation in future bliss as long as the body remained undecomposed. Pouqueville tells us that, "Whenever a bishop or priest excommunicates a person, he adds—'after death, let not thy body have power to dissolve.' In consequence of this malediction the following remarks are found in a manuscript of the Church of St. Sophia at Thessalonica. (1.) Whoever has been laid under any curse, or received any injunction of his dead parents that he has not fulfilled, after his death the fore part of his body remains entire. (2.) Whoever has been the object of any anathema appears yellow after his death, and his fingers are shrivelled. (3.) Whoever appears white has been excommunicated by the Divine laws. (4.) Whoever appears black has been excommunicated by a bishop."[19] In this way it is possible to discover the crime for which, and even the person by whom the judgment has been pronounced. One horrid result of this ghastly superstition is the custom which is prevalent among the Greeks of Salonica, and the Bulgarians in the interior of European Turkey, and probably also over a much wider area, of disinterring the dead after a year spent in the grave, in order to judge by the state of their bodies whether their souls are in heaven or hell.[20]

Corresponding to the power of excommunication here described is that of absolution, which resides in the

---

[19] Pouqueville, 'Voyage de la Grèce,' vi. p. 153.
[20] Mackenzie's and Irby's 'Travels in Turkey in Europe,' i. pp. 15, 130.

person who pronounced the ban, or in some superior ecclesiastical authority. Of the effect of this in decomposing the now rigid corpse there are numerous and well-authenticated instances; indeed we may learn from some of them that the most direct evidence, though established on excellent authority, must be disregarded, if we are to arrive at the truth in matters of this kind, where on the one side credulity is prevalent, and on the other the temptation to the employment of fraud is strong.

The following curious story is told by Martin Crusius relating to the times just subsequent to the taking of Constantinople. There were about the court of Mahomet II. a number of men learned in Greek and Arabic literature, who had investigated a variety of points connected with the Christian faith. By them the Sultan was informed that the bodies of persons excommunicated by the Greek clergy do not decompose, and when he inquired whether the effect of absolution was to dissolve them, he was answered in the affirmative. On this he sent orders to Maximus, the patriarch of that time, to produce a case, by which the truth of the statement might be tested. The Patriarch convened his clergy in great trepidation, and after long deliberation they discovered that a woman had been excommunicated by the preceding Patriarch for the commission of grievous sins. When they had found her grave they proceeded to open it, and lo! the corpse was entire, black, and swollen like a drum. When the news of this reached the Sultan, he despatched some of his officers to possess themselves of the body, which they did, and deposited it in a safe place. On an appointed day the liturgy was celebrated over it, and the Patriarch recited the absolution in the presence of the officers; on which, wonderful to relate, the bones were immediately heard to rattle as they fell apart in the coffin, and at the same time, the narrator says, the woman's soul was also freed from the punishment to which it was doomed. The courtiers at once ran and informed the Sultan, who was astonished at the miracle, and exclaimed, "Of a surety the Christian religion is true."[21]

This story sounds somewhat apocryphal; let us take some others which are better attested. Leo Allatius, who, as a Roman Catholic, was not likely to be too favourable to the Greek clergy, informs us that he was told by Athanasius, Metropolitan of Imbros, an honourable man, and one whose word was to be trusted, that on one occasion, being earnestly entreated to pronounce the absolution over a number of corpses that had long remained undecomposed, he had consented to do so, and before the recitation was concluded they all fell away into ashes.[22] Ricaut also, in his account of the state of the Greek Church, written in 1678, relates a very similar occurrence, to which he appends the following remark by way of explanation:—"This story I should not have judged worth relating, but that I heard it from the mouth of a grave person, who says that his own eyes were witnesses thereof."[23]

The ways in which these spirits may be laid are very various, and the priests in particular have great power in exorcising them. The people of Sphakia in Crete, Mr. Pashley tells us, believe that the ravages committed by these night-wanderers used in former times to be far more frequent than they are at the present day; and that they are become comparatively rare, solely in consequence of the increased zeal and skill possessed by the members of the sacerdotal order. From the same writer we learn

[21] Crusii, 'Turco-Græcia,' pp. 133-6. [22] Leo Allatius, as above, p. 151.
[23] 'Present State of the Greek and Armenian Churches,' p. 282.

that the people of Hydra attribute their present freedom from them solely to the exertions of their bishop, who has laid them all in Santorin, where on the desert isle they now exist in great numbers, and wander about, rolling stones down the slopes towards the sea, "as may be heard by any one who passes near, in a kaik, during the night."[24] In Mytilene the bones of those who will not lie quiet in their graves are transported to a small adjacent island, where they are reinterred. This is an effectual bar to all future vagaries, for the vampire cannot cross salt-water.[25] Probably, however, the method of proceeding adopted by a Thessalian bishop, mentioned by Colonel Leake, was the most certain in its operation. "The metropolitan bishop of Larissa lately informed me," he says, "that when metropolitan of Grevená, he once received advice of a papas having disinterred two bodies, and thrown them into the Haliacmon, on pretence of their being Vrukolakas. Upon being summoned before the bishop the priest confessed the fact, and asserted in justification that a report prevailed of a large animal having been seen to issue, accompanied with flames, out of the grave in which the two bodies had been buried. The bishop began by obliging the priest to pay him 250 piastres (his holiness did not add that he made over the money to the poor); he then sent for a scissors to cut off the priest's beard, but was satisfied with frightening him. By then publishing throughout the diocese that any similar offence would be punished with double the fine, and certain loss of station, the bishop effectually quieted all the vampires of his episcopal province."[26]

In the town of Perlepe, between Monastir and Kiuprili,

[24] Pashley, ii. p. 202.   [25] Newton, i. p. 213.
[26] Leake's 'Northern Greece,' iv. p. 216.

there exists the extraordinary phenomenon of a number of families which are regarded as being the offspring of *vrykolakas*, and as having the power of laying the wandering spirits to which they are thus related. They keep their art very dark, and only practise it in secret; but nevertheless their fame is so widely spread that persons who are in need of such deliverance are accustomed to send for them from other cities. In ordinary life they are avoided by the whole world.[27] Mr. Newton mentions the following mode of exorcism, which was employed in Rhodes on a woman who had returned to earth as a *vrykolaka*. "The priest of the village laid on the ground one of the dead woman's shifts, over the neck of which he walked, held up by two men, for fear the vampire should seize him. While in this position he read verses from the New Testament, till the shift swelled up and split. When this rent takes place the evil spirit is supposed to escape through the opening."[28] Saturday is the day of the week on which the exorcism ought by rights to take place, because the spirit then rests in his tomb; and if he is out on his rambles when the ceremony takes place, it is unavailing. In most parts of the country, as the vampire is regarded as only a night-wanderer, he has to be caught during the night between Friday and Saturday;[29] but in some places, where he is believed to roam abroad by day as well, the whole of Saturday is allotted to him for repose, and consequently is suitable for his capture.[30] The idea here is probably connected with the observance of the Sabbath. Other ways are mentioned by which the evil influence may be averted or modified. The *vrykolaka* stands in mortal fear of the sign of the cross. Thus a shepherd, who was

---

[27] Hahn, 'Albanesische Studien,' i. p. 163.
[28] Newton, i. p. 212.
[29] Hahn, as above.
[30] Pashley, ii. p. 201.

driven by a storm to take refuge in the arched sepulchre of a man who had undergone this change, and had committed great ravages among the inhabitants of the neighbouring districts, determined to pass the night there, without being aware on whose tomb it was that he was about to lie. Fortunately for him, however, before going to sleep he took off his arms, and placed them crosswise on the stone at his head. In consequence of this, the vampire was unable to rise, and when he besought the shepherd to remove them, he refused to do so until the spectre had sworn by his winding-sheet (the only oath by which these creatures are bound) that he would do him no harm.[31] Amongst the Wallachians a variety of preventive measures are taken at the time of burial to hinder the dead from returning to earth. Thus, for instance, a long nail is driven through the skull, and the thorny stem of a wild rosebush is laid upon the body, that its winding-sheet may become entangled with it, if it attempts to rise.[32]

When exorcising, and other methods of laying the spirit are of no avail, the last resource is to burn the body. To this the Greeks resort only in great extremities, as they have a great horror of cremation. The Turks, however, as I have already said, have no such scruples about burning the body of a Christian. The Wallachian custom is to drive a stake through the heart, a mode of procedure which used also to be practised in Western Europe. The following narrative, given by Tournefort at the commencement of the last century, of the state of things of which he was an eye-witness in the island of Myconos, describes the various processes that were resorted to on such an occasion, and at the

---

[31] Pashley, ii. p. 198.   [32] Schott, p. 298.

same time presents a lively picture of the horror with which the popular mind is inspired by this superstition:—

"The man whose story we are going to relate was a peasant of Mycone, naturally ill-natured and quarrelsome: this is a circumstance to be taken notice of in such cases. He was murdered in the fields, nobody knew how, or by whom. Two days after his being buried in a chapel in the town, it was noised about that he was seen to walk in the night with great haste; that he tumbled about people's goods, put out their lamps, griped them behind, and a thousand other monkey tricks. At first the story was received with laughter; but the thing was looked upon to be serious when the better sort of people began to complain of it. The Papas themselves gave credit to the fact, and no doubt had their reasons for so doing: masses must be said, to be sure; but for all this the peasant drove his old trade, and heeded nothing they could do. After divers meetings of the chief people of the city, of priests and monks, it was gravely concluded that 'twas necessary, in consequence of some musty ceremonial, to wait till nine days after the interment should be expired.

"On the tenth day they said one mass in the chapel where the body was laid, in order to drive out the demon which they imagined was got into it. After mass they took up the body, and got everything ready for pulling out its heart."

Then follows a somewhat anatomical description of the process, which was performed by the butcher of the town, and was followed, of course, by such an escape of gases, as required a large amount of incense to neutralize it.

"Their imagination, struck with the spectacle before them, grew full of visions. It came into their noddles that a thick smoke arose out of the body; we durst not say 'twas the smoke of the incense. They were incessantly bawling out *Vroucolakas* in the chapel and place before it. This is the name they give to these pretended *redivivi*. The noise bellowed through the streets, and it seemed to be a name invented on purpose to rend the roof of the chapel. Several there present averred that the wretch's blood was extremely red; the butcher swore the body was still warm: whence they concluded that the deceased was a very ill man for not being thoroughly dead, or, in plain terms, for suffering himself to be reanimated by Old Nick, which is the notion they have of a Vroucolakas. They then roared out the name in a stupendous manner. Just at this time came in a flock of people, loudly protesting they plainly perceived the body was not grown stiff when it was carried from the fields to church to be buried, and that consequently it was a true Vroucolakas; which word was still the burden of the song."

The next step was to burn the dead man's heart on the sea-shore;

" but this execution did not make him a bit more tractable. He went on with his racket more furiously than ever. He was accused of beating folks in the night, breaking down doors and even roofs of houses, clattering windows, tearing clothes, emptying bottles and vessels. 'Twas the most thirsty devil! I believe he did not spare anybody but the consul in whose house we lodged. Nothing could be more miserable than the condition of the island: all the inhabitants seemed frightened out of their senses; the wisest among them were stricken like the rest. 'Twas an epidemical disease of the brain, as dangerous and infectious as the madness of dogs. Whole families quitted their houses, and brought their tent-beds from the farthest part of the town into the public place, there to spend the night. They were every instant complaining of some new insult; nothing was to be heard but sighs and groans at the approach of night; the better sort of people retired into the country."

Processions and sprinklings of holy water were then tried for several days running, but with no better success.

" One day, as they were hard at this work, after having stuck I know not how many naked swords over the grave of the corpse, which they took up three or four times a day for any man's whim, an Albanese, that happened to be at Mycone, took upon him to say, with a voice of authority, that it was to the last degree ridiculous to make use of the swords of Christians in a case like this. " Can you not conceive, blind as ye are," says he, " that the handle of these swords, being made like a cross, hinders the devil from coming out of the body? Why do you not rather take the Turkish sabres?" The advice of this learned man had no effect: the Vroucolakas was incorrigible, and all the inhabitants were in a strange consternation. They knew not now what saint to call upon, when, of a sudden, with one voice, as if they had given each other the hint, they fell to bawling out all through the city that it was intolerable to wait any longer; that the only way left was to burn the Vroucolakas entire; that, after so doing, let the devil lurk in it if he could; that 'twas better to have recourse to this extremity than to have the island totally deserted; and, indeed, whole families began to pack up, in order to return to Syra or Tinos. The magistrates therefore ordered the *Vroucolakas* to be carried to the point of the island St. George, where they prepared a great pile with pitch and tar, for fear the wood, as dry as it was, should not burn fast enough of itself. What they had before left of this miserable carcase was thrown into the fire, and consumed presently: 'twas on the first of January, 1701. We saw the flame as we returned from Delos; it might justly be called a bonfire of joy, since after this no more complaints were heard against the *Vroucolakas*. They said that the devil had now met with his match, and some ballads were made to turn him into ridicule." [23]

The vampires that have been hitherto spoken of have been almost entirely of a malignant character. That there are others, however, which are quite innoxious, is shown by the following poem, which is entitled 'The Vurkolakas.' It is one of the finest of all the Romaic ballads; and at the same time it possesses an additional interest, from our being able to trace in it very clearly the way in which these songs are brought to perfection. Many versions of it exist, and most of them, like the one given in Fauriel's collection, are of inferior merit; but in these we can trace the process of refinement, through which both the subject and the expressions have passed, and which has ultimately brought it to the state in which we find it here.

"THE SPECTRE.

"A mother had nine gallant sons, and one beloved daughter,
One only daughter, dearly prized, the darling of her bosom;
For twelve long years she suffered not the sun to rest upon her,
But washed her at the fall of night, and combed her locks ere daybreak;
And while the stars were still on high ranged them in dainty tresses.

---

[23] Tournefort, 'Voyage to the Levant' (Eng. Trans.), i. p. 103, foll.

Now, when there came an embassy from a far distant country,
And sought to lead her as a bride into the land of strangers,
Eight of her brothers were averse, but Constantine approved it:
'Nay, send her, mother mine,' he said, 'into the land of strangers;
To the far country I frequent, where I am wont to travel;
So I shall gain a resting-place, a comfortable mansion.'
'Prudent you are, my son,' she said; 'but your advice is evil.
What if fell sickness visit me, or gloomy death approach me,
Or joy or sorrow be my lot, who then shall fetch her for me?'
He sware to her by God on high and by the holy martyrs,
That if dark death should visit her, or sickness fall upon her,
Or joy or sorrow be her lot, he would go forth to fetch her.
And so she sent her Areté into the land of strangers.
But when there came a season fraught with pestilential sickness,
And they were struck with fell disease, and the nine brothers perished,
Then, like a bulrush in the plain, the mother sat deserted.
At all their tombs she beat her breast and raised her lamentation;
But when she came to Constantine's she lifted up the gravestone,
And 'Rise,' she cried, 'my Constantine; I need my darling daughter.
Didst thou not swear by God on high, and by the holy martyrs,
When joy or grief became my lot, that thou would'st go to fetch her?'

"Lo! from his tomb—there, where he lay—her invocation raised him.
He rode upon the stormy cloud, the stars bedecked his bridle;
His escort was the shining moon; and thus he went to fetch her.
Before his flight the mountains rose, and disappeared behind him,
Till he beheld her where she combed her tresses in the moonlight.
Then from afar he called to her, and from afar he hailed her:
'Come with me; come, my Areté! our mother calleth for thee.'
'Alas!' she answered, 'brother dear, at such an hour as this is!
Say, is thy summons one of joy? shall I put on my jewels?
Or, if 'tis gloomy, tell me so, and I'll not change my garments.'
'Come with me; come, my Areté! wait not to change thy garments!'

"Then as they passed along the road, accomplishing their journey,
The birds began to sing aloud, and this was what they uttered:
'O strange! a spirit of the dead leading a lovely lady!'
'O listen, Constantine,' she said, 'to what the birds are singing:
'"O strange! a spirit of the dead leading a lovely lady!"'
'Regard them not, the silly birds; regard them not, my sister.'
So they passed on; but, as they passed, again the birds were singing,

'O wondrous pitiable sight! O mystery of sadness,
To see the spirits of the dead walking beside the living!'
'O listen, listen, brother dear, to what the birds are singing:
'"Behold the spirits of the dead walking beside the living!"'
'Regard them not, poor birds,' he said; 'regard them not, my sister.'
'Alas! I fear thee, brother mine! thy garments smell of incense.'
''Tis naught,' he said; 'on yestereve we worshipped at the altar,
And there the priest, in passing by, fumed me with clouds of incense.'
And once again, as they passed on, yet other birds were singing,
'Almighty God! thine hand it is this miracle that worketh,
To send a spirit of the dead to lead this lovely lady.'
She heard the voices as they spake, and her heart sank within her.
'O listen, listen, Constantine, to what the birds are singing!
Say, where are now thy golden hair, and flowing fair moustaches?'
'A wasting sickness fell on me, near to the grave it brought me;
'Twas then I lost my golden hair and flowing fair moustaches.'

"They came; and lo! the door was closed, the bolt was drawn before it,
And the barred windows, one and all, with spiders' webs were tangled.
'Open,' he cried, 'my mother dear! behold, I bring thy daughter!'
'If thou beest Charon,[34] come not here; I have no other children.
My Areté, unhappy one! lodges far off with strangers.'
'Open, my mother! tarry not; 'tis Constantine that speaketh!
Did I not swear by God on high, and by the holy martyrs,
When joy or grief became thy lot, that I would go to fetch her?'
She rose; and as she reached the door the mother's soul departed!"[35]

---

[34] Charon is the modern Greek personification of Death.
[35] Passow, 'Carm. Pop.' No. 517.

CHAPTER XXII.

PELION.

Meeting-point of Ossa and Pelion — Ruins at Skiti — Chain of Fortresses on Pelion — Ferocious Dogs — Ali Pasha regretted — Woods of Polydendron — Ruins of Casthanæa — Keramidi — Ascent to the Ridge — Places named from Trees — Plain and Mountains of Thessaly — Intelligent Greek farmer — Views of the Eastern Coast — Zagora — Hospitable Reception — Prosperity of the Townships of Magnesia — Romaic and Neo-Hellenic Languages — Dancing the Romaika — Inferior Position of Women.

An hour's ride to the south-east of Aghia brought us to a large dry river-bed, spanned by a Turkish bridge of one wide arch. Hard by this there is a little plain, forming the innermost recess of the plain of Aghia; and this, together with a valley that runs in from it to the east, form the line of demarcation between Ossa and Pelion. Still, the separation of the two mountains is not complete, so that, as Herodotus observes, they mingle their roots.[1] We proceeded to mount the steep ascent of Pelion at this point, and in another hour reached some extensive ruins near the village of Skiti. These occupy a fine position in the form of a triangle, the apex of which faces north-east, overlooking the valley just mentioned; and as the rocks to the west are nearly precipitous, and those to the east quite so, there seem to have been hardly any defences on those sides. The principal line of walls follows the southern slope at the base of the triangle, having a tower in the middle: in the neighbourhood of this tower a good deal of brick is mixed with the stones and materials of the walls, but there is not a

[1] συμμίσγοντα τὰς ὑπωρέας ἀλλήλοισι.—Herod. vii. 129.

trace of Hellenic work in any part; the blocks which M. Mézières refers to that date are merely large stones, not squared.² Still it is highly probable that these mediæval walls occupy the place of an Hellenic city, and one, moreover, of some consideration on account of the importance of its situation. For though the plain of Aghia lies several hundred feet above the sea, and the lowest part of the ridge where Pelion and Ossa meet is considerably above it, yet Mézières is right in regarding this as a practicable entrance to Thessaly, especially as the coast in this part retires sufficiently to make a tolerable roadstead. Now the city of Melibœa is described by Livy as situated on the roots of Ossa, facing Thessaly, and in a position suitable to command Demetrias,³—that is to say (as that place was on the Pagasetic Gulf), fitted to enable an enemy to turn it and cut off its communications with Macedonia by Tempe. Now this very closely corresponds to our present position; and the likelihood of their identity is increased by Strabo's statement that Melibœa was situated in the gulf between Ossa and Pelion,⁴ which shows that it commanded the sea, though it could not have been on it, in consequence of Livy's remark that it faced Thessaly.

The ruins of which we have been speaking, though not Hellenic, are of considerable interest in connection with

---

² The opinion that Leake has propounded with regard to this and similar buildings in Thessaly, that they are Hellenic, though their walls are composed of small stones and mortar, mixed with broken tiles ('Northern Greece,' iv. 406), is very strange, and so entirely opposed to what we find elsewhere, that, notwithstanding the authority of the great geographer, we cannot hesitate to reject it. That the Greeks of Thessaly did not build differently from those of Southern Greece, we shall see when we examine the ruins of Casthanæa.

³ Livy, xliv. 13. See Mézières, p. 77. No great difficulty need arise from the historian's saying that it was on Ossa, as it was so near the point of junction of the two mountains.   ⁴ Strabo, ix. 5, § 22.

a very ingenious suggestion of M. Mézières with regard to a number of fortresses of similar construction which he discovered on Pelion. According to him, there are remains of eight such buildings in different parts of the mountain, forming a chain of defensive works, five of which are on the western side—three overlooking the Pagasetic Gulf, and two the plain of Thessaly—while the remaining three command the coast of the Ægean.[5] From these he thinks we may conclude that, at one period of the Byzantine empire, Pelion was converted into a stronghold to command the neighbouring plain, and to serve as a refuge in case of invasion. Unfortunately we have no further evidence to confirm this by no means unreasonable hypothesis, nor any information on the subject from the Byzantine historians; and in my own mind the doubt rises whether some of these remains are not Venetian, for the name of Veneto, which belongs to one of the villages of the eastern coast, seems to show that that people at one time had settlements here; and with this we may connect the tradition mentioned by Mézières,[6] that the inland parts of Pelion were first colonized by Greeks from Euboea, who fled from the tyranny of a Venetian duke. Still there is a completeness in the other view which makes us unwilling to pick holes in it, though we must be content to leave it as a doubtful question.

My visit to Skiti explained to me another point connected with ancient remains, and that, too, in rather a summary manner. My companion had gone off to another part of the site, leaving me to examine the southern wall in company with a young Turkish *Zaptié*, or policeman, whose services as guide in this difficult

---

[5] Mézières, pp. 30, 90.   [6] 'Le Mont Pélion,' p. 34.

country we had obtained from the Mudir at Aghia. As we approached the central tower we were suddenly attacked by half-a-dozen ferocious dogs, which had been asleep within it together with the shepherd, and on hearing our footsteps rushed upon us at close quarters and were all round us in an instant. Fortunately, however, in Greece, wherever there are dogs there are also stones to throw at them (χερμάδια), and with these missiles we defended ourselves until the shepherd came and called them off. Now I had seen the remark in books of travels, and had often noticed it myself, that the dogs in these countries are fiercest in the neighbourhood of the old buildings, but had failed until now to discover the cause, which is evidently this—that these places are the most convenient shelter for shepherds, and that, consequently, near there you are most likely to come upon their dogs unexpectedly, and to be the object of a violent sally.

In two hours from Skiti we reached the village of Polydendron, which, like all the others along the coasts of Ossa and Pelion, is built at a considerable elevation above the sea, having a little port or *scala* by the shore ; the reason apparently being that, at the time of their foundation, the fear of pirates was too great to allow of a position being chosen nearer to the water. At present the fears of the inhabitants seem to proceed from another quarter. M. Mézières remarks that the people of this part of Thessaly look back with regret to the days of Ali Pasha, which they regard as "the good old times." It is curious to compare this with the cry which Leake mentions as being universal at that period—" Albania has ruined us" (Μὰς ἐχάλασεν ἡ 'Αλβανητία). *Now,* "the peasantry remember that he did not crush them with taxation."[7]

[7] Mézières, p. 75.

Then we hear that the "hungry plunderers" whom he and his sons introduced "devour at least the provisions of the poor peasantry, if they carry their extortion no further."[8] The principle of *laudatio temporis acti* may in part account for this; but at the same time the French writer is probably not far wrong in saying, "Le gouvernement turc est parvenu à faire regretter Ali-pacha : c'est sa plus sanglante condamnation, et la meilleure réponse à ceux qui croient ce gouvernement encore en progrès."[9]

The name of Polydendron, or "Woodlands," seemed at first sight hardly applicable to this place, as there are few trees about it, though our attention was attracted by two weeping willows in the village itself—very rare trees in Turkey. Very soon, however, we discovered, to our cost, the derivation of it, for about a mile further on there commences an extensive region of matted woods, composed of an inextricable maze of arbutus, catalpa, juniper, and branching heather. After dining by a spring on the seashore, in the midst of flowering myrtle-bushes, we set to work to force our way through this tangled region, which extended for five or six miles along the mountainside, descending the steep slopes even to the water's edge. The greater part of the way had to be accomplished on foot, for the branches had so overgrown the path, and were so closely entwined, as to threaten every moment to sweep you off your horse, if you attempted to ride : but as we climbed up and down the sides of the dells, the views were exquisite, and closely resembled those of the finest parts of Athos, except that the round though imposing outline of the summit of Pelion to the south did not make up for the glorious pyramidal peak.

---

[8] Leake, 'Northern Greece,' p. 325. [9] Mézières, p. 76.

Many of the rocks in this part are composed of *verd antique;* and all through this district of Olympus, Ossa, and Pelion there are numerous peculiarities in the strata which would be highly interesting to a geologist. At last we emerged into more open ground, over which our route continued, until we were within half an hour of the village of Keramidi, our resting-place for the night: here we sent on our dragoman, and leaving our horses on the hill-side in the care of the young Zaptié, descended ourselves towards the sea, to examine some important Hellenic remains, which we had seen from above. We were already aware from Mézières' book that they were to be found somewhere in this position; but our muleteer, Yani, who had accompanied us from Letochoro, and for a Greek was an extraordinary dolt, looked upon it as a sort of divination on our part, and exclaimed in astonishment, "Who in the name of fortune told them that there were ruins here?" (ποῖος διάβολος τοὺς εἶπε, πῶς ἦτον παλαιόκαστρο ἐδῶ;)

When, after a rough scramble through the prickly bushes, we reached these walls, we found that they occupied a hill running from west to east, and projecting into the sea, to which it descends in rugged precipices at its eastern end. Both on the northern and southern side also the ground falls very abruptly, so that the only part of the site where the slope is comparatively easy is that to the west, where the acropolis is placed on a sort of hump, which is the highest point in the whole position. The walls by which this was defended are very strong, being composed of large blocks of stone put together without mortar, in the finest style of Hellenic masonry: that which separates the acropolis from the lower town is especially fine, and has round towers at its extremities, near the northern of which was the gate of entrance. In this wall the blocks are fitted together in places at irregular angles, or, where interstices have been left, they are filled in with smaller stones, nicely inserted, but the courses of masonry are parallel throughout the whole of the ruins.[10] The best-preserved piece of wall is that at the south-west angle, where seven courses remain, overlooking a deep hollow in the rock, which seems to have been excavated for a quarry. In the other walls there are square towers at intervals, but the northern and eastern sides seem to have been hardly at all defended : the entire circuit cannot be much more than half a mile; the whole position very closely resembles that of Rhamnus, in the neighbourhood of Marathon. Near the precipices at the eastern end there is a small modern chapel of St. George, built of stones rudely put together. There can be little doubt that this place represents the Casthanæa of antiquity, which is the only town besides Melibœa mentioned by Strabo[11] as being on this side of Pelion, and is named by Herodotus in his account of the destruction of Xerxes' fleet on this ironbound coast.[12] The

---

[10] It has generally been thought that, where the blocks are built in irregularly in the manner just described, it is a sign of greater antiquity in the construction, and walls of this sort have received the names of Cyclopian and Pelasgic; but in reality the difference between these walls and others of more regular construction is to be attributed almost entirely to the nature of the stone employed, the polygonal form of the blocks in the one case being caused by the hard and jagged character so often found in the limestone of Greece, which was at once irregular in its fracture and difficult of working; while in the other case, the regular cleavage of the rocks, or their less intractable material, suggested the use of rectangular blocks, arranged in horizontal courses. (See Ross, 'Inselreisen,' i. 15.) No doubt in later times, when there were greater facilities for stone-cutting, the parallel system became all but universal; but it was in use also at the earliest period, as is shown from its constant occurrence in the oldest walls in Italy, where the material is mostly a soft tufa, and whose builders, be it remembered, were of the same Pelasgic race to which the earliest Greek architecture is to be referred.

[11] Strabo, ix. 5, § 22. [12] Herod., vii. 183, 184.

learned men of Zagora, a town which is a day's journey further south, put in a claim in favour of that place, and support it by the abundance of chestnut trees in that neighbourhood, while none are found near Keramidi. It is true that the name of that tree is connected with Casthanæa; in fact we are informed that it was actually derived from it:[13] but my impression is that so great a

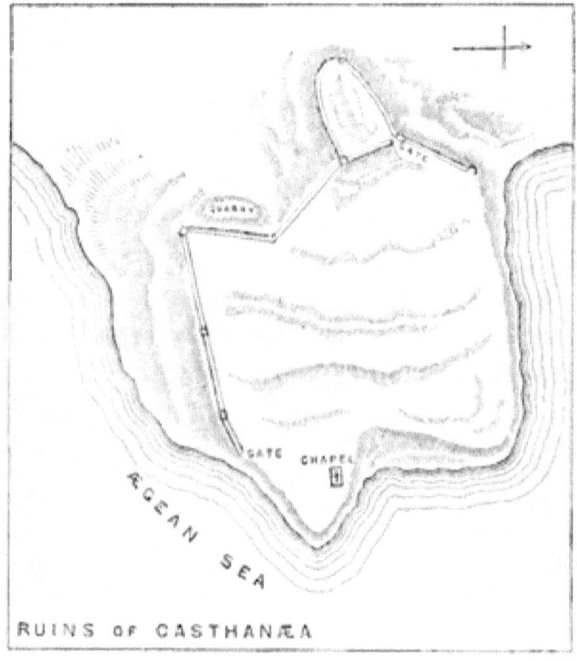

Ruins of Casthanæa.

change had passed over the vegetation of these parts between ancient and modern times, that arguments based on any exact correspondence between them are of little value. To this subject I shall have before long to return.

At Keramidi we took up our abode with the lame

---

[13] 'Etymol. Magn.,' quoted by Leake, 'N. Greece,' iv. 382.

schoolmaster of the place, who presided over a school of sixty pupils. The door of his house was closed at night with a wooden bar drawn across it, and run into holes in the two sides, exactly the μοχλός of the ancients; and in one of the walls was the following monumental inscription:—

ΔΑΜΟΚΡΙΤΑ
ΑΙΝΕΙΟΥ
ΠΟΛΥΑΙΝΩ
ΑΙΝΕΙΟΥ.

We were strongly advised to take a boat from here, and go by sea to Zagora, for owing to the rugged and tangled character of the country, of which we had already had some experience, this is the regular mode of communication among all these villages; so that the paths are hardly known, and it is no easy matter to find a guide. This, however, would have defeated our object, which was to explore the interior; and accordingly, as our young Zaptié was forced to return to Aghia, we hired the services of a Greek farmer called Constantine, a most intelligent man, and an excellent specimen of his class. The people of Keramidi, like all the inhabitants of Pelion, complained bitterly of the extortionate taxes laid upon them, having to pay irregular exactions (δοσήματα) in addition to the regular imposts (δέκατα); but for all that, they seemed from various signs to be fairly well off; and among others, from the large sums of money which guides demand in payment, quite out of proportion to what is usual elsewhere.

Instead of keeping near to the coast of the Ægean, we now struck up westward to the ridge of Pelion, through a region of oak forests, interspersed with a few elms. The highest point here was 2000 feet above the sea, but the mountain rises rapidly towards the south, in the direction of the summit, so that when we again crossed it, later in the day, we found we had ascended to 4200 feet. The ground was excessively stony, and we came to no water for five hours, a sufficient cause to have prevented this district from being inhabited in ancient times. During a good part of the way we were on the west side of the ridge, and at last we descended a little distance on that side to the village of Kerasia, where there is a copious spring. The name of this place, "the Cherries," reminds us how numerous throughout Greece and Turkey are the localities called from trees. In every country, as might naturally be expected, this is one not infrequent source of nomenclature; but here it prevails in an extraordinary manner, proving, amongst other things, the completely rural character of the population, whether Hellenic or Slavonic, at the time when these names were given. Thus on Olympus we have noticed Carya "the nuts," and Leftocarya, "the hazels," and near Rapsani there is a village called Krania, "the cornels." On Pelion, besides Kerasia and Polydendron, there is Kissos, "the ivy," near Zagora; and among Slavonic names we find Bukova, Bukovina, "the beeches," above Monastir, Arachova, "the nuts," on the side of Parnassus, Navarino,[14] "the maples," Vervena, "the willows," and others. If the reader will take the trouble to examine the modern district of Laconia in Kiepert's map of Greece, he will find that out of a very moderate number of names the following have a similar derivation:—Castania, from the chestnut, Platanaki and

---

[14] Other etymologies have been suggested for this name; 'Αβαρίνων, as having been the supposed seat of an Avar colony (Finlay's 'Greece under the Romans,' p. 418); 'Ομερίνο, from the fact that its castle was built in the thirteenth century by Nicolas de Saint Omer (Clark's 'Peloponnesus,' p. 225). The derivation from *javor*, the Slavonic for "maple-tree," is perhaps the more probable, as following the more usual analogy. In all cases the Greek *ν* has been prefixed.

Platana from the plane-tree, Kyparissi from the cypress, Apidia from the pear, Achladokambos from the wild pear, Sykia from the fig, Valanidia from the valonea oak, Myrtia from the myrtle, Daphni from the laurel, Arachova from the nut, Kydonia from the quince; and others bear the names of plants, as Marathonisi of the fennel, and Phlomochori of the mullein. In many places, however, we find that though the name remains, the tree itself is no longer found, which is the case at Kerasia; and with regard to this place, M. Mézières remarks that it is customary among the Greeks, in accordance with some tradition or other, to give the name of "the cherries," which he had found also high up on Mount Helicon, to certain elevations. This, however, is singularly unlikely, and we may feel certain that the tree has at some time or other existed at the place where the name is found: only it is important to remark that such names are frequently given rather on account of the rarity than the abundance of the tree; that rarity giving it its distinctive character, and consequent significance as a name, and at the same time accounting for its easily becoming extinct.

The views on this side of the ridge of Pelion were extremely striking, and as they comprised all the south-eastern part of the Thessalian plain, formed an excellent supplement to that from the south-western angle of Mount Olympus. The peak of Ossa here forms a majestic object, rising far above all the intervening mountains, and amply justifying its name of "the look-out place," "the watch-tower;"[15] to the north-west appear the heights to the south of Larissa, together with the spur that descends from the Cambunian chain; to the south-west the Karadagh, from which also a ridge runs off to the north, and

---

[15] Ὄσσα is probably derived from ὕσσομαι, as σκοπιά, σκόπελος from σκέπτομαι. See Curtius, 'Grundzüge der Griechischen Etymologie,' ii. 51.

behind it the huge masses of Othrys. At our feet lay the long lake of Karla, anciently the lake of Bœbe, a most unpleasant-looking piece of water, marshy and full of reeds; when the sun shone upon it from the west, it looked like a slimy bituminous swamp, and anything rather than "fair-flowing," which is the epithet Euripides bestows upon it.[16] The parched slopes, too, on the further side of it had a strange tint of warm yellow, which reminded my companion of the Arabian desert. The ground descends towards it on all sides, so that it has no escape for its waters, which are fed by the neighbouring streams, and by the overflow of the Peneius, which, as I have already said, discharges its surplus waters into the lake Nessonis, and from thence into this lower piece of water. As a watershed intervenes between this and Volo, we could understand how this part of the Thessalian plain appeared comparatively little below us, notwithstanding our elevation above the sea. In the lake there is a valuable fishery, but two years ago the water dried up, and this was for the time destroyed; now, however, it has recommenced, the fish having returned from the streams that feed it. Again, to the south were the Gulf of Volo (Pagaseticus sinus), the broken heights of Eubœa, and in the far distance the lofty summits of Parnassus. The peak of Pelion appeared like a horn at the further end of its long ridge.

To this ridge we again mounted, and then continued for some hours to ascend along it, or rather through a trough-like depression between two ridges, which were everywhere covered with beech-forests. Our guide Constantine now became very communicative, and shewed the usual keen inquisitiveness of his race. He asked us

---

[16] καλλίναον Βοιβίαν λίμναν.—Eur. 'Alc.,' 589.

questions about various European languages, about the words that corresponded in them, and the possibility of an original connexion between them—a subject which would hardly be found to interest many English farmers. When we spoke to him of the ruins we had visited on the day before, he told us the tradition at Keramidi was that they had been destroyed by the Pope of Rome—a story which shows, like that which I have before mentioned as existing on Athos, how widespread the idea of some such invasion is in the Ægean. As to vampires and similar apparitions, *he* did not believe in them, but most of the people in this country did: every village along the coast of Pelion, he said, was haunted by a *vrykolaka* of its own, with one exception—and that was the village of Sklithro, which was situated high up on the mountain-side between Polydendron and Keramidi. From him also we learned that the oak in these parts is called δένδρον, as being the tree *par excellence;*[17] this is one among many instances in modern Greek of a generic name being used specifically for the most marked class that falls under it, as "beast" in English commonly signifies "bullock." The modern Greek for a horse, ἄλογον (as being the most important "irrational animal"), is a well-known instance; ῥιζάρι also, the word for madder, is similarly the root *par excellence*. Colonel Leake somewhere mentions that in the islands, where there are no horses, donkeys are called πράγματα, "facts," or "possessions:" but in reality throughout the district which we have been traversing the word πρᾶγμα (pronounced *pramma*) is applied to all beasts of burden, whether horses, mules, or asses, as they are the most important possession. In the Cretan

---

[17] The same took place in ancient Greek, where δρῦς, the word for "oak," originally signified "tree."—Max Müller's 'Lectures on the Science of Language,' 2nd Series, p. 219.

dialect κτῆμα, we are told, is used in the same way for a horse;[18] and κτῆνος similarly signified a beast of burden in Hellenistic Greek.

After descending some distance on the other side of the ridge, we at last caught sight of the south-eastern district of the mountain, stretching away in a succession of undulating slopes of fertile land dotted with white houses, beneath which there is a steep descent all along to the shoreless sea, forming the "harbourless coast of Pelion"[19] —the terrible Sepias, where so many vessels of Xerxes' fleet were dashed to pieces. Beyond are seen the beautiful forms of the islands of Sciathos, Scopelos, and the others which run off from the extremity of the promontory. We were benighted before reaching Zagora, and had to scramble by the light of the stars down a very rough path, and afterwards down a still rougher paved staircase, which led into the town. It was the eve of the festival of the Transfiguration, for the celebration of which Zagora is the head-quarters for all this neighbourhood; and in consequence of this the place was very full, and accommodation difficult to procure: so we were lodged in what was called the church-house, a place of meeting for the authorities, both civil and ecclesiastical, containing a couple of rooms.

Early the next morning our toilettes were hardly completed when our humble apartment was entered by a deputation of the chief men of the place, mostly dressed in Frank costume, except for the fez cap, who came to express their deep regret that two English gentlemen (δύο εὐγενεῖς Ἄγγλοι) should have been consigned to so unsuitable an abode, and explained that it was only the

---

[18] See Lord Strangford's valuable appendix on that dialect in Spratt's 'Crete,' i. p. 362.
[19] ἀκτὰν ἀλίμενον Πηλίου.—Eurip. 'Alc.,' 595.

lateness of our arrival that rendered such a mischance possible. When we protested that we were in better quarters than what we were accustomed to, they renewed their apologies, and begged us to give them an opportunity of making some amends by spending the festival with them, and taking up our abode in one of their families. To the first part of their request we gladly acceded, but we valued our liberty too much readily to give in to the latter. However, the chief medical man, Dr. Fronimos, pressed us so urgently to come to his house, which was allowed by common agreement to be the best in the place, that we found we could not persist further in our refusal without giving offence, and went accordingly. We had no reason to repent of doing so, for it was superior to any dwelling-house that we had seen out of the great cities of Turkey—spacious, airy, and free from furniture, except the hard cushions with which the divans were covered. With true Oriental politeness we were left to ourselves during the greater part of the day on a balcony commanding a delicious view of the isles of Greece, and the grand peak of Athos, here nearly 80 miles distant, which could not fail to occupy our thoughts, as it was the anniversary of the day when, four years before, we had stood on its summit. The hottest part of the day all along this coast is from seven till ten in the morning, when the sun is in the east; after that time the sea-breeze rises, which they call the Hellespontine, and which no doubt corresponds to the Etesian winds of classical authors, as those were said to blow during August.

The village of Zagora is situated in the midst of a forest of chestnuts on the steep mountain-side; and though with the adjoining village of Pori it numbers 850 houses, and therefore seems rather to deserve the name of a town, yet it is really a large village, all the houses being separate and surrounded by trees, so that it spreads irregularly over a very wide space of ground, ranging from twelve to fifteen hundred feet above the sea. Its exports are silk and oil, of the former of which 3500 okes, and of the latter 6000 okes and upwards are annually exported. All the commerce passes by way of Volo at the western foot of Pelion, for the *scala* below the village is considered unsafe as a landing place, as well it may, if, as Leake conjectures, it represents Ipni,[20] the scene of the greatest destruction of Xerxes' fleet. The remoteness of the place is implied by the name Zagora, a Slavonic word signifying "behind the mountain," which is found in several other parts of Turkey applied to districts in a corresponding position, such, for instance, as that *behind* Pindus, to the north-west of the Zygos pass and Metzovo; and that *behind* the Balkan, relatively to Bulgaria—the district of Sophia and Philippopolis. But the wealth of the place depends mainly on its long standing commercial connection with foreign countries. Formerly, indeed, its inhabitants were much occupied in the manufacture of the cloth called *skutia*, and that which was made here is said to have been finer than any produced elsewhere; this, however, like all the other manufactures of this district, has now come to an end. The mine also of mixed lead and silver, which some years ago was worked here by a company of Constantinople merchants and others, and managed by Englishmen residing on the spot, has now been given up, not in consequence of a failure in the supply of ore, but from want of capital to carry it on. But a large number of persons belonging to the families of Zagora have been for many generations,

---

[20] Leake, 'N. Greece,' iv. 383.

and are still, engaged in trade abroad, including many of the wealthiest Greek merchants in Constantinople, Smyrna, and Alexandria, and in many of the ports of western Europe. Of these latter few return, as they become enamoured of European life; but notwithstanding this, they do not forget the place of their birth; thus there are two schools at Zagora—one a higher or Hellenic school for boys, where they are taught Latin and the ancient Greek classics, the other for the education of girls—both of which are wholly supported by a rich merchant in England. But that the majority of them still regard this as their home, we had ocular demonstration in the multitude of Smyrniote and Alexandrian merchants and shopkeepers in Frank dress, of whom the place was full at the time of our visit; these had escaped from the cholera that was then raging in those cities, and had taken the opportunity of revisiting their families.

The flourishing condition, not only of this village, but of all the twenty-four townships of which the district of Pelion, or Magnesia, as it is still called, is composed, is a truly delightful and cheering sight in the midst of the many miseries and general mismanagement of this unhappy country. It is confessedly the most prosperous district of the whole of Turkey, and its inhabitants, though they complain of the burdens laid upon them, are quite aware that they enjoy far more freedom than the other *rayahs*. Throughout the whole of it, except in one or two places at its western foot, there is not a single Turkish village; and though guards or policemen of that race are quartered on them, yet at Zagora the captain of these is a Christian. The origin of this prosperity is probably to be referred in the first instance to the fact that a large number of the villages are *vacouf*, i.e. Mahometan church property, and consequently exempt from taxation and inalienable; the rest has been done by its inaccessible position, by the fertility of the soil, and above all, by the enterprise and industry of the Greek population, unimpeded by interference and fear of spoliation.

In the course of the morning we were taken to the church of the Transfiguration, to be present at the service. This building, which from its dedication is the great centre of attraction on this day, dates from the 12th century, and was built, we were told, by one of the Comneni: in plan it follows the ordinary type of Byzantine churches, except that the body of the church has a bay more than usual. The frescoes are ancient—much more so than those in the monastery of St. Demetrius, and rich in their general effect; on one of the columns I noticed that the decorations of the capital were evidently imitated from ancient masks, a sign of a more original and less ecclesiastical taste than is to be seen elsewhere. It is a very dark building, though the effect of this was lessened by the number of tapers, which, as is customary on festivals, were lighted in all parts of it. Near the church precincts were placed a number of stalls, where objects were exposed for sale, mainly consisting of silks, calicoes, and trumpery ornaments, which formed a great contrast to those of some of the women, who wore large necklaces of coins, or belts with shields of filigree work in silver gilt. After we returned from the service, we had a second visit from the chief man of the place, Mr. Gallopulos, a portly and rather pompous old gentleman, who had headed the deputation that waited on us in the morning. He is considered a very wealthy man, for besides considerable business as a merchant, in which capacity he has dealings with firms in England and elsewhere, he possesses lands at Volo and in Greece. When we enquired of him about the library established here at the end of the last century by one of their fellow townsmen called John Prinko, he answered that there was little in it of any value, and that the few German and French books it contained are not read : indeed, he led us to believe that no one at Zagora speaks anything but Greek, though we were afterwards informed at Volo that he himself speaks a little French, and that Dr. Fronimos can, at all events, *write* Italian. He evidently prided himself on the purity of his Neo-Hellenic, as the regenerated language of modern Greece is called to distinguish it from the vulgar Romaic ; and as the last improvement in pronunciation he mentioned, that in literary circles at Athens a distinction is beginning to be introduced between the *υ* and the other *e* sounds, with which it has hitherto been confused, so that it should be pronounced like *u* in French, which was probably the sound it represented in ancient Greek.

The progress which has been made in improving this language by eliminating all Turkish words, expurgating barbarous forms, and enriching the vocabulary, is truly wonderful ; and there can be little doubt that it will ultimately be made an admirable vehicle of thought, from its plastic nature, its capability of forming new words to express new ideas by derivation and composition, and the resources it has to draw from in the ancient tongue. Especially will this be the case, if while the system of inflections is in great measure restored to its original purity, the analytic form of syntax which it has in common with other modern European languages, is not sacrificed. But these advantages have their corresponding drawbacks, arising especially from the rapidity with which the change has taken place. To pass over the affectation of numerous phrases and forms of speech evidently adopted from French modes of expression, the fact is, that what has been effected elsewhere by literature in the course of several centuries in the way of establishing a standard and pruning away deformities has here been done in thirty or forty years. The consequence is, that members of the same race are already speaking almost different languages; and if the process goes on with the same rapidity, there is a danger of a line of demarcation being drawn between the upper and lower classes almost as strong as that which separated our Norman and Saxon forefathers. An intelligent French merchant, who had resided many years in Thessaly, and other provinces where Greek is spoken, and could converse fluently with the natives, having learnt the language by ear, assured me that he had great difficulty in understanding a Greek newspaper; and we ourselves, when at a later period of our tour we entered the kingdom of Greece and conversed with Athenians and others of the educated class, were surprised to find that some of the most ordinary words, which we had heard in daily use, were unintelligible to them, and that we were forced to replace them by synonyms drawn from the ancient language. The older generation, it is true, from having once known something of the Romaic, generally understand it, though they ignore and despise it; but as to the younger Athenians, I really believe they would have greater difficulty in communicating with the Greek peasants in Turkey, and even in their own country, than foreigners with a superficial knowledge of the language acquired in the mountain districts.[21]

[21] It may be well to mention, for the information of future travellers, that the ordinary modern Greek grammars are, for the reasons given above, of comparatively little use to persons visiting the interior. The best suggestion that I can make on this subject is to procure a grammar published

In the evening we went to see the dancing, which had been kept up, to the sound of a drum and two clarionets, ever since the morning service. The scene of it was an open space behind the church, with a circular paved area, like that of a threshing floor. The accessaries of this were admirably picturesque : all round it rose enormous plane-trees, and in front was a view of the sea and islands in contrast with the cupola of the church and one grand tapering cypress, while behind, where the ground is steep, the women were arranged all together on the slopes in a semicircle, like the spectators in an ancient theatre. The men stood round the area, and the whole assemblage was very large, multitudes having come by sea from the villages of Pelion for a long way round. It was consequently an excellent opportunity for studying the physiognomy of the people of the district : many of the men were tall of stature, and a few of the women were good looking, but the Bulgarian cast of face decidedly predominated amongst them, though a fair number had Greek features, and the dark Greek eye was far the most common. The dresses were unusually commonplace, as the men wore the dark-coloured baggy Hydriote trowsers ; in fact, the only thing worth noticing in the whole multitude was the effect produced by the large number of crimson fezes. The performance, too, was disappointing ; I never saw the Romaika worse danced : on this occasion it quite deserved the name Byron gives it of "a dull roundabout," which certainly is not applicable to it as it is performed in parts of the Morea, where a graceful serpentine movement is often introduced, and the hands of those who join in it are linked very elegantly.

before, or shortly after, 1830, when the old Romaic was universally spoken ; the best are in German. In the way of a dictionary, I have found none as useful as the small German and Mod. Greek Lexicon, published by Tauchnitz.

As a national dance, however, there is always this to be said for it, that it has the advantage of being easy, and of allowing a large number to take part. The company is formed into a ring, broken at one point, with the ends overlapping, so as to admit of indefinite extension, even to the formation of an inner coil; a step or two forward, and then a step or two backward, followed by one to the side, keeps the whole ring in movement and circulating. The largest number we saw dancing together was six-and-thirty; of these but few were women, and those of that sex who did join, performed their part in a most business-like manner, and with a sobriety of deportment worthy of a very solemn function indeed. A few of the non-residents in Frank costume took part, the most conspicuous among them being an old gentleman in a hat three sizes too big for him, and wearing a general appearance of seediness, as if he had lately turned out from Holywell Street: he danced most vigorously. For some time our host, the doctor, led the dance, which he did with grace and dignity. The men seemed thoroughly to enjoy it, and relinquished the scene unwillingly at nightfall.

We returned home to supper, where we were waited on, not by the servants, but by the mistress of the house and her children, our host partaking with us at the table. This is the custom in these parts, and betrays the fact that even in this civilised community the Oriental feeling with regard to the inferiority of women still prevails. We found other proofs of the same thing. At Makrinitza, the most important village on the western side of Pelion, the attempt was made some time ago to establish a school for girls, but it met with so great opposition that it was ultimately abandoned, the inhabitants being afraid of their women being much educated. Again, at

Zagora and elsewhere the value set on women is so small, that when a female child is born there is weeping and mourning in the house, so that with regard to one sex they actually realise the custom of the "Thracian wives of yore."[22] A girl is considered an expense and unremunerative; for if she marries she must have a dowry, and if she does not, she becomes a permanent burden to her family. In consequence, the parents endeavour to get their daughters married extremely early, and when they reach nineteen they are looked upon as almost past the marrying age.

[22] Herod., v. 4.

CHAPTER XXIII.

PELION (*continued*).

Superb View of Athos — Ancient and Modern Vegetation of Pelion — Ascent to the Summit — View from it — Cave of Chiron the Centaur — Descent to Portaria — Site of Iolcos — Hill of Goritza — Ruins of Demetrias — Volo — Ill-treatment of Bulgarian Labourers — Excursion to Milies — No Turks on Pelion — Homer and Virgil on "the Giants' Mountains" — Olives and Oil — Library and School of Anthimos Gazes — Quarantine.

It would be hard to find any sight on earth more beautiful than that which greeted our eyes as we rose just after daybreak the next morning. Above the eastern horizon were spread clear translucent spaces of that "daffodil sky," with which our observant Laureate has familiarized us, but which nevertheless is rarely seen; and in the midst of them rose from the sea an object "phantom-fair," yet sharply cut, distinct and real — the gigantic watch-tower of "lonely Athos." We both thought we had never seen so superb a mountain. The clearness of its outline and the delicacy of the tints soon faded away when the sun's disk appeared above the water; but it was one of those scenes which, once beheld, are never forgotten!

After bidding our hospitable friends an affectionate farewell, we left Zagora, with the captain of the guards for our guide, and mounted slowly by a steep path towards the summit of the mountain, close to which the road to Volo passes. Below us to the south lay the scattered villages of Kissos and Anelios, and beyond them the beautiful slopes and gorges of the further part of

Magnesia, with the islands, and in the distance Scyros, once the home of Achilles. The ridge of Pelion, with all the ground for several hundred feet below it, is thickly covered with beech-trees; but there is not to be seen on the whole mountain either a pine-tree to build a ship Argo,[1] or an ash to make a spear for Achilles.[2] The vegetation of Pelion, however, is too interesting a subject to be dismissed in a few words, as we have here the means of comparing the ancient and modern condition of a district in this respect.

There has come down to us a geographical notice of this mountain,[3] the greater part of which is taken up with a descriptio of the trees and plants which grow there—a sufficient proof that the Homeric epithet of Pelion, "quivering with foliage" (εἰνοσίφυλλον) was as applicable then as it is now. From this we learn that, while there was a great variety of different kinds, especially of fruit trees, the most abundant trees were the beech, the silver fir, two sorts of maple, the cypress, and the juniper.[4] Those that I noticed as growing there at the present time are the oak, the beech, the chestnut, the plane, the elm, the ilex, the olive, the weeping willow, and the cypress; and amongst shrubs the arbutus, the myrtle, and the laurel. Of these the beech is far the most common, though the oak is found in considerable quantities on the northern part of the ridge, and the chestnut

---

[1] μηδ' ἐν νάπαισι Πηλίου πεσεῖν ποτε
τμηθεῖσα πεύκη.—Eur. 'Med.,' 3, 4.

[2] Πηλιάδα μελίην.—Hom. 'Il.,' xvi. 143.

[3] This has been erroneously attributed to Dicæarchus. See Müller's Prolegomena to the 'Geographi Græci minores' (p. lii.), among whose works the 'Descriptio Montis Pelii' is given.

[4] πλείστην δ' ὀξύην ἔχει καὶ ἐλάτην, σφένδαμνόν τε καὶ ζυγίαν, ἔτι δὲ κυπάρισσον καὶ κέδρον. For the proof that these trees correspond to the modern names given above, see Dr. Daubeny's 'Trees and Shrubs of the Ancients,' pp. 7, 20, 36, 42, 46.

covers the mountain sides at Zagora. Among the rest, the weeping willow probably does not grow wild; nor the cypress either, notwithstanding that it is spoken of as indigenous in classical times, for I have never seen it, except as a cultivated tree, in any part of Greece or Turkey, though in burial-grounds and in the neighbourhood of buildings it grows luxuriantly.[5] Neither the maple nor the juniper came under my observation; but the most noticeable fact was the entire absence of any kind of fir or pine.

After this comparison we are almost forced to the conclusion that, whatever the cause may be, the change which has passed over the face of the country in respect of its trees is very great, and generally, that arguments based on the supposed correspondence of these in certain districts, in ancient and modern times, are of little value.

The summit of Pelion lies at a little distance to the west of the ridge, and not more than 300 feet above the path which we were following. We left our horses near a spring at its foot, and, following our guide, scrambled up its steep side through the beechwoods, from which at last the bare rocky peak emerges. I had expected little from the view, for it seemed from below as if the southern heights of the mountain would shut out a great part of it; as we ascended, however, these soon sank down below us, and the more prominent parts of the ridge close to us only excluded a portion of the Ægean and the mainland of Chalcidice, leaving moreover an opening formed by a depression, through which Athos was visible. When, at last, we stood on the top, great was my surprise and delight to behold a panorama of

---

[5] Col. Leake has remarked the same thing.—'Northern Greece,' iii. 397, *note*.

extraordinary magnificence, certainly superior to that from Olympus, and unequalled, to my mind, by any view in Greece, except that from the summit of Parnassus. The most distinguishing feature is the land-locked gulf of Volo, a most beautiful basin, which lies below you in its whole expanse, with headlands at intervals running out into it, extending from the small town and closed harbour of Volo at its head to the narrow outlet at its southern end, where the chain of Pelion, turning at right angles to its axis, throws out a long projection with a narrow isthmus and broken outline. At the further end of this is the isle of Trikeri, lying off the town of the same name, the ancient Aphetæ, the scene of the departure of the Argonauts. To enumerate in succession the more distant objects, beginning from the south-east—next to the islands, which we had already seen during our ascent from Zagora, appeared the long coastline of Eubœa, here seen in profile, with the conical peak of Mount Dirphe, its highest summit, which rises behind Chalcis, forming a conspicuous object; then a succession of inland pieces of water—the straits of Artemisium, part of the Euboic, and part of the Maliac gulf; beyond the last-named of which lies Mount Œta, in itself no despicable range, but here altogether dwarfed by the huge mass of Parnassus, which towers over everything near it; next Othrys, a noble chain, with the plain of Larissa, now much obscured by haze, and Lake Bœbe in all its swampy length; further to the north the sharp summit of Ossa, overtopped by the broad but less imposing Olympus, and finally a portion of the Thermaic gulf, and that part of Chalcidice which projects into its waters. It was indeed a scene of surpassing magnificence; but independently of its grandeur it would be hard to find another with so many and so remarkable features of historical interest. The height of the mountain was given by the barometer as 5515 feet, somewhat higher than the number 5310 feet, as determined by the Admiralty survey. The fact that it is now always called Pelion, and that its Slavonic name Plessidi (from *pletch*, a mountain ridge), which is given in all the maps, and must at one time have quite superseded the ancient name, is never heard, while that of Kissavo is universally used for Ossa, must be taken as a further proof of the revival of classical sympathies and Hellenic feeling amongst the communities by which its flanks are inhabited.

When we had sated our eyes with gazing on this glorious spectacle, the next object of our search was the cave of Chiron the Centaur, the tutor of Achilles. Leake[6] would place this some distance to the south of this point, in a depression between this and the next summit, where there is, he says, a fine cavern. Its position, however, is very exactly pointed out in the following curious passage of the 'Description of Mount Pelion:' "On the highest peak of the mountain there is a cavern called the Cave of Chiron, and a temple of Zeus Acræus,[7] to which at the time of the rising of the Dog Star, when the heat is most intense, the young men of highest rank among the citizens (*i.e.* of Demetrias), after having been selected in the presence of the priest, ascend clothed in new fleeces of thick wool, on account of the extreme cold on the mountain." From this it is evident that the temple occupied the summit on which we were standing,

---

[6] 'Northern Greece,' iv. 384.

[7] In the MSS. of the 'Descriptio Montis Pelii,' the reading is Διὸς 'Ακταίου, and not 'Ακραίου; but in two inscriptions existing in the district of Pelion, and copied by M. Mézières, the title of the god is 'Ακραῖος. This, taken together with the suitableness of the appellation, is amply sufficient to justify us in altering the text. *See* the inscriptions in Mézières, pp. 117, 118.

and that the cave of Chiron was near it. Now there is a cavern about 30 feet below this point, though it is now ruined by a fall of rock, and nothing of it is visible but the hole by which it was entered from above. Our guide said it had long been in this state, but that old people, who remembered it before it was destroyed, had told him that there was a flight of steps leading down into the earth, and that the sides of the cave were covered in some places with paintings, from which we may gather that it was probably a hermit's retreat. This cavern is regarded by the people of Zagora as the cave of Chiron, and its position corresponds so closely to that just described, that there can be hardly any doubt of its identity.

When we had returned to the place where we left our horses, and after breakfasting by the spring had indulged in a comfortable *siesta* (proving thereby that *we* did not feel the cold, as the young Greeks of former days are described to have done), we commenced the descent on the other side. As the view towards the west is not visible from the base of the peak, those of our company who had not mounted with us had not as yet come in sight of it; in consequence of which we had the opportunity of witnessing a very remarkable effect produced by it upon one of their number. This was our muleteer Yani, whom I have already described as a great dolt, and a most unimpressible being. When, however, the lake-like expanse of the gulf suddenly burst upon him, lying beneath his feet, he appeared quite bewildered, and gave vent to his feelings in a loud and prolonged exclamation of open-mouthed astonishment. It was a complete triumph of nature over man! From this point a rough path conducted us in zigzags down the bare mountain side to the village of Portaria, which, notwithstanding the length of the descent, is more than 2000 feet above the sea. We may here remark with regard to Olympus, Ossa, and Pelion, what is true also of Athos, that their eastern slopes have always more natural vegetation than the western, probably in consequence of the former having a more northerly aspect, and therefore being less scorched by the reflection of the burning sun from the rocks. The important villages of Portaria and Makrinitza, which like Zagora contain many wealthy inhabitants and good houses, are divided from one another by a deep ravine, and have a picturesque appearance from their steep position, the numerous cypresses with which they are planted, and the separate arrangement of the houses, though they are not scattered as widely here as in some other places. The manufactures from which they formerly derived a large revenue now no longer exist, but silk and oil are still exported. The name of Portaria is probably derived from the Latin *porta* (which is naturalized in modern Greek), signifying the gate or entrance, as it commands the high road of traffic to and from the other side of Pelion,[8] while that of Makrinitza, another instance of a mixed Greek and Bulgarian compound, is evidently taken from the long line of buildings with which it clambers up the rocky mountain-side.

The following morning we continued our descent by a path even steeper than that of the previous day. Our ultimate destination was the port of Volo, which lay directly below us; when, however, we began to approach the plain, we followed a track leading to a point further to the south, in order to examine the extensive ruins called Goritza. On the way we passed a hill which bears the name of Episkopi, with a ruined church crowning

---

[8] *See* Barth, 'Reise,' p. 218.

its summit. It is a singularly fine position for an old Hellenic city, as it lies at some little distance from the sea, so as to be safe from the attacks of pirates, and projects from the side of Pelion, to which it is joined by a lower ridge, with a rocky summit for the acropolis. It hardly admits of doubt that this was the site of Jason's far famed city, Iolcos, the centre of many a poetic legend, for it is the only position in this neighbourhood which would have been chosen for a town in the early times, except the hill of Goritza, which was certainly occupied by Demetrias. Strabo says that the ruins of the place (for it was destroyed long before his time) were situated above Demetrias, at seven stades distance from it; such, at least, seems to be the meaning of a somewhat obscure passage;[9] and this would just correspond to its situation relatively to Goritza, for it is about a mile from that place, and nearer to the mountain. As might be expected in a site so long deserted, there are no Hellenic remains. As we descend southward from this, we pass through the village of Vlakho-makhala, the stream that pours down through which is the ancient Anaurus, which, according to the same geographer, flowed close to Demetrias.[10] This torrent was the scene of a romantic incident in the life of Jason, which has been prettily told by Apollonius.[11] On its banks one day, as he was returning from the chase, "when all the mountains and lofty peaks were sprinkled with snow, and the torrents descending from them swept roaring along in their courses," Juno met him in the

---

[9] ἡ δ' Ἰωλκὸς κατέσκαπται μὲν ἐκ παλαιοῦ. τῆς δὲ Δημητριάδος ἑπτὰ σταδίους ὑπέρκειται τῆς θαλάττης Ἰωλκός. (Kramer suggests καὶ τῆς θαλάττης.) Strabo, ix. 5, § 15.

[10] πλησίον δὲ τῆς Δημητριάδος ὁ Ἄναυρος ῥεῖ.—Strabo, ut supra.

[11] 'Argonautica,' iii. 66 seq., and i. 8.

guise of a helpless aged woman, and he took pity on her and bare her on his shoulders through the raging flood : but in so doing he lost one of his sandals, and thus, when he appeared before his uncle Pelias, he was recognised by him as the one-sandalled man who was destined to overthrow him. The state of the torrents here described is very similar to what must often be seen at the present day ; for the consuls at Volo complain that the streams which come down from Makrinitza, Portaria, and this village, form for themselves a number of ill-defined channels, and reach the sea at a variety of points, thus rendering the place unhealthy.

The hill of Goritza is a conspicuous mass of rocky ground, which projects into the sea between two small plains—that of Volo on one side, and that of Lekhonia on the other — and bars the communication between them. No site could be more appropriate for a large city, which should command the approach to Thessaly in this direction ; and consequently Demetrias, which was built here by the great Demetrius Poliorcetes, was spoken of as one of the three "fetters of Greece," Chalcis and Corinth being the other two. It occupied a level on the sea-face of the hill formed by the spreading of the root, that here runs off from Pelion, beyond which the rocks descend three or four hundred feet in a broad mass to the water's edge. Formerly it was surrounded by walls, but its northern side is additionally strengthened by the formation of the ground ; for in that direction it is crowned by a steep ridge some 500 yards in length, extending from W.S.W. to E.N.E., with an outer slope, steep and rapid, to the plain of Volo. This ridge is the most interesting part of the site, being generally a mere narrow *arête* of rock with a wall all along it ; but at the north-east end it widens slightly, so as to admit of buildings. There does not seem to have been any enclosed acropolis, but only forts; one at the south-west end, and two at a certain interval from one another at the north-east. In one part of the *arête* there are two openings or breaks close to one another, and possibly there may at one time have been a small gate of entrance in this part, as the rocks have been cut away. At the north-eastern end of the ridge is an ancient enclosure of blocks even with the ground, containing a small roofless chapel, on the north side of which is seen the mouth of a cistern, while to the east an irregular hollow has been excavated in the rock, 22 feet square and 12 feet deep, in which there is a descent to the mouth of a well. This enclosure is the scene of an annual miracle on Easter Sunday, when the well-mouth, from being dry, becomes full of water, and continues so during the day—no doubt in consequence of some subterraneous communication with the cistern. From this place there is an ascent of 160 yards to the highest point of the ridge, where the *arête* commences. A great part of the circuit of the ancient walls below may be traced, the conspicuous tower which rises above the sea at the southern extremity of the town being the only modern addition. The walls in the neighbourhood of this tower are formed of large blocks laid in regular courses, but generally the masonry is composed of small blocks roughly put together. On the level there are traces of an underground aqueduct, and of lines of streets measuring 15 feet across.

From the ruins of Demetrias we descended to the *scala* of Volo, or the New Warehouses (κινούρια μαγάζια), as it is called, to distinguish it from the Kastro or Turkish walled town, and from Old Volo, a Greek village lying at the foot of the mountain beneath Portaria. This is the residence of consuls and merchants, and we found it greatly increased in size since we last saw it in 1853; a change which is easily explained by the fact that, until lately, the building of houses was interdicted here by the Turkish Government. Our countryman, Mr. Borrill, the Italian consul, with great kindness accommodated us with a room in his house, though he was destitute of servants in consequence of the fever, which had been very prevalent in the lowlands during this summer. We could not help remarking, as we descended from the highlands of Magnesia, where every one seemed robust, and the number of old men was especially noticeable, that the first person whom we met in the plain was suffering from this malady. From our windows we could see on the opposite side of the harbour the piers of a Roman aqueduct, though neither lofty nor of fine construction, stretching inland for some distance: these belong to the ruins of Pagasæ. Close to the shore, on the nearer side of these, was situated the *lazzaretto*, which had just been the scene of a lamentable piece of Turkish mismanagement. When the cholera broke out in Constantinople, there were a number of Bulgarian labourers in the place, who gain a livelihood there as masons, carriers, &c., returning to their homes in Western Turkey for three months during the summer. Two thousand of these applied to the Sultan to be allowed to leave the place, and he with great good feeling provided them with a steamer to convey them free of expense to the nearest port. Accordingly they were sent to Volo, where they were to perform quarantine. But as no notice had been given and no preparations made, the *lazzaretto* proved wholly inadequate for their reception, and this vast multitude was turned out into the neighbouring country, which was hot and destitute of shade, without provisions of any kind. Here they continued in the greatest misery, until the consuls protested, and they were removed to the island of Trikeri at the mouth of the gulf, where they had at least the shelter of trees. But even there so little care was taken of them, that some almost died of hunger and the cholera broke out amongst them. About the time of our arrival they were released. As usual, though the original conception was good, the execution was as bad as could be. It is one of a thousand examples which show the constitutional weakness of the "sick man;" and this, far more than absolute depravity, is what we have to complain of in him. In saying this I do not mean to acquit the Ottoman Government of more serious charges; recent circumstances have shown that it can connive at, if not encourage, acts of a most disgraceful character. But while, on the whole, we may believe it to be well intentioned, we cannot help condemning it of utter, irremediable incompetency. No doubt there are men of great ability among its statesmen, and some real and important reforms have been carried out of late years, such as the conversion of the *kaimés*. But it is useless to dwell on these, and speak of them as "progress," when the general decay is proceeding at the same time at a far more rapid rate; so that all improvements are but the mending and patching of a rotten and threadbare system. Skilful diplomacy, together with the adoption of a certain number of European ideas, may for a time blind other nations to the presence of a festering mass of corruption in their neighbourhood, but to those who have the misfortune to live under it we may be sure that the Turkish rule will be productive of nothing but depression and misery.

The ports of Turkey, and Volo among the number, were now so thoroughly compromised by the cholera, that we had no hope of escape without a long quarantine; and the accounts we had received in the course of our journey of that established on the Greek frontier were so exaggerated, that it was almost a relief to find that the necessary time was eleven days. There was nothing for it, therefore, but to make up our minds to endure this penance. Before its commencement, however, we determined to have one more expedition on the sides of Pelion, and on the day following our arrival at Volo we started for Milies, a village about six hours distant to the south-east, and the next in importance after those we had already seen. Our road at first passed round the rocky surface of the hill of Goritza, underneath the position of the ancient city, and then entered the fertile plain of Lekhonia, which is covered with maize, and vines, and olives. In one place the figs were being dried in the sun, having been laid on a framework of reeds, extended on upright poles about ten feet from the ground. The *kharidji*, from whom we hired our mules, was a good-humoured, lively, vain Greek, far superior to most men of his calling, and a great contrast to the dullard Yani. We had not gone far with him before he let us know that he had accompanied Prince Arthur to the summit of Pelion, and helped to raise the pile of stones which now exists there as a memorial of his visit. Two things with regard to His Royal Highness seemed to have made a peculiarly strong impression on his mind: the first was that during the ascent the Prince had continually said to him "*covàn, covàn,*" the meaning of which expression was an enigma to us at first; but when he went on to say that the Turkish guard who accompanied them echoed it by "*chabouk, chabouk*" (quick, quick), we perceived that it was "go on, go on:" the other was that, on parting, the Prince had given him an English pound, which I told him was a princely *baksheesh*.

At a point somewhat less than half-way to Milies two extensive villages are seen on the mountain-side; first, that of St. Lawrence, and then that of St. George. Between them lies a deep ravine, and near the plain at its mouth is the town of Lekhonia, the only one besides Volo in the district of Pelion that contains Turkish inhabitants and is surmounted by a minaret. The reason of this is to be found in the position of those two places in the plains, for the Turks have never ventured to settle in these mountains. Formerly the population of Lekhonia was entirely Turkish; now, however, the majority are Greeks; and even the Turks regularly use the Greek language. On the height above appear the ruins of a mediæval castle with a conspicuous tower, one of those which, according to Mézières, formed the Byzantine circuit of defence. Still higher in the same direction rises the second summit of Pelion, between which and the one we ascended there is a considerable depression; while, on the other side of it, the ridge of the mountain slopes downwards, and continues comparatively low until it reaches the sea.

The depression here mentioned is that which Dr. Holland [12] noticed as suited to explain that part of the fable of the Giants' attempt to scale heaven which relates to Ossa and Pelion, since the upper part of the latter mountain at that point seems to form a broad foundation, on which another mountain of more conical form, like Ossa, might rest. The remark is a very ingenious one, but, unfortunately, it applies not to Homer's description, but to that of Virgil, who, in imitating his great predecessor, has inverted the order in which the three mountains were piled up, and certainly cannot be supposed to have im-

[12] 'Travels,' ii. 95.

proved upon him, since he places Olympus, the most massive of all, at the apex. As Mr. Gladstone has observed,[13] Homer's description is in conformity with the proportionate heights of the mountains; among which Olympus is the highest, Ossa the next, Pelion the least; the pyramidal structure, which satisfies the eye, being thus adopted. If the appearance of the mountains, however, were taken into account, and not their size, it would be far more fitting for Ossa to rest on Pelion.

From the summit just mentioned a great spur descends beyond Lekhonia and projects into the sea, its lower slopes being covered with a vast extent of olive-yards. There can be little doubt that this is the "greatest and most thickly-wooded root" spoken of in the 'Description of Mount Pelion' as being seven stades by sea and twenty by land from Demetrias.[14] Mézières also is probably right in regarding the stream which flows down through the gorge as being the Crausindon mentioned in that work, of which it is said that it irrigates the cultivated lands at the foot of Pelion. At this time of year both this stream and those near Volo, though in the upper part of their courses they have a considerable supply of water, fail notwithstanding to reach the sea, being drawn off, like the rivers of Attica, for purposes of irrigation.

---

[13] 'Homeric Studies,' iii. 528. The two passages are—

"Ὄσσαν ἐπ' Οὐλύμπῳ μέμασαν θέμεν, αὐτὰρ ἐπ' Ὄσσῃ
Πήλιον εἰνοσίφυλλον."—Hom. 'Od.' xi. 315.

" Ter sunt conati imponere Pelio Ossam
Scilicet, atque Ossæ frondosum involvere Olympum."—
Virg. 'Georg.,' i. 281.

[14] Τοῦ δ' ὄρους ἡ μεγίστη καὶ λασιωτάτη ῥίζα τῆς πόλεως κατὰ μὲν πλοῦν ζ' ἀπέχει στάδια, πεζῇ δὲ κ'.

When we had passed this spur, we began to ascend into the mountains, and at last, after crossing a pretty stone bridge of one arch, thrown across the bed of a winter torrent, arrived at Milies, a village of 350 houses, which lies more than half-way up the mountain-side, withdrawn from the sea, but commanding a lovely view of the lower part of the gulf and of the mountains. Conspicuous among these were Parnassus, and the lofty peak of Veluki, the ancient Tymphrestus, the Liubatrin of this end of the Pindus chain, which stands at the head of the valley of the Spercheius, and forms a marked object in most of the panoramic views of North Greece. Here, as in the other villages of Pelion, the houses are scattered over the steep slopes; among them were growing a great variety of fruits—walnuts, chesnuts, figs, grapes, pears, mulberries, and peaches. The people of the place were excessively inquisitive, and as I was resting myself under some umbrageous plane-trees in an open space in the village, a crowd formed round me, and left me no peace from the interminable string of questions they addressed to me on every conceivable subject, until one of the chief men came up and removed them to a respectful distance, where they remained listening to our conversation. From him I learnt that 300,000 okes of oil are exported yearly from this place; but in making this estimate they are obliged to take the average of two years, as it is only in alternate years that the olive-crop is good. This year we hardly saw any olives on the trees in any part of Magnesia; and the same thing seems to be the case in other places in the Levant, for Mr. Newton[15] notices it as occurring in the island of Mitylene. According to him the explanation is to be

---

[15] 'Discoveries in the Levant,' i. p. 82.

found in the custom of beating the fruit off the trees, instead of gathering it ; so that the olive does not seem to relish the treatment which the old maxim prescribes for walnut-trees, together with spaniels, and some other beings higher in the order of creation. At Zagora we were told that the whole amount of oil exported in a very favourable year from all the twenty-four townships was three million okes ; this year it would not exceed five hundred thousand. A considerable quantity of silk is also produced at Milies, and at Lekhonia there is a silk manufactory. When I inquired about a place called Kokkali, *i.e.* "The Bones," he answered that it is situated two hours further up towards the ridge of the mountain, and that, as the name implies, a great quantity of bones are found all about the neighbourhood of it ; but neither he nor any one there present knew of any tradition connected with them. It has been said that a great battle was once fought there in ancient times,[16] but the story has probably been invented to account for the name.

While we were at dinner in the khan, we were visited by a gentleman from Alexandria, a native of this place, who possesses a house at the top of the village, to which he comes for six weeks once in two years. He rushed into our apartment, and in a series of sentences in broken English, far too ludicrous to be repeated, begged us with great importunity to come and pass the night with him. It was with great difficulty that we made our excuses, engaging at the same time to take a cup of coffee with him in the morning. When we repaired to his house at the appointed time, we found it very inferior to many we had seen at Portaria and Zagora, but it was delightful to

---

[16] Mézières, p. 35.

find a man so enraptured with his home and its rural pleasures, to which, at least in his present humour, he would willingly have retired for life. In his company we visited the public library, which was founded at the beginning of this century by a priest called Gregory, and subsequently increased by another native of Milies, Anthimos Gazes, a man of great acquirements, who had travelled much in Western Europe. These two persons, together with a brother of Gregory called Daniel, conceived the idea of founding a college here, as a resort for students from various parts of Greece; and this excellent project, which might have given a great start to education, was on the point of being set on foot when, like the similar one of Eugenius Bulgaris on Athos, it fell through owing to the jealousy felt by the neighbouring townships. The library is said to have been damaged by the Turks during the War of Independence, and is now in disorder, and apparently not used. It consists of Greek and Latin classics, and some ecclesiastical works, together with books on mathematics, and a good many in French and German, though there were hardly any of the classic authors of those languages among them. The editions were mostly worthless, but there were two fine ones of Ovid and Virgil, with verse translations in Greek opposite. The same building contains also an Hellenic school with about thirty scholars, of which Gazes was the founder, and in which he taught till the end of his life. His name, together with that of the two brothers, is still remembered and reverenced, as it deserves to be.

We had now reached the term of our explorations, and in the course of the day returned by the same route to Volo. There we found a boat, and crossed during the night to the quarantine station, in the little island of St.

Nicolas, near Mintzela, on the Greek frontier. After eleven days, not unpleasantly spent there in reading and writing, we were picked up by a small Greek steamer, which calls at that place weekly, and, passing through the Euripus at Chalcis, were landed at the Piræus on the 7th of September.

CHAPTER XXIV.

THESSALY AND METEORA.

Journey in 1853 — Departure from Salonica — Approach to Thessaly — Larissa — A Grandee of the Old School — Plains of Larissa and Tricala — Tricala — Modern Adaptations of Ancient Names — Rocks of Meteora — The Great Monastery — Admittance Refused — Monastery of St. Stephen — Remarkable View over the Plain of Thessaly — Fear of Robbers — Monastery of Barlaam — Ascent by the Rope and Net — Ballad on the Abbot of Barlaam.

I WILL now ask my readers to suppose themselves once more at Salonica, and to carry themselves back in time to the summer of 1853, shortly before the beginning of the Russian war. During the spring of that year we had travelled over a great part of Greece, accompanied this time by a third fellow-traveller, and with Nicola Kombotegra of Athens for our dragoman. We had now just returned from our first visit to Athos, and were preparing to ride across to Corfu by a lower route than those which have already been described, passing through Thessaly and Epirus by the cities of Larissa and Yanina. It is this journey which I now propose to describe. The first part of the way lies through country with which we are more or less acquainted, and consequently we may pass over it quite rapidly, until we reach the western extremity of the Thessalian plain; from thence we may start afresh, and penetrate once more into Albania.

At nightfall on the 10th of June we dropped out of the harbour of Salonica. As we receded from the shore, the city presented an enchanting sight, for each of the tall minarets which rise from among its houses was encircled by a ring of glittering lamps, as is the custom during the fast of the Ramazan, when the Mahometan population indemnify themselves by the festivities of the night for the enforced abstinence of the day. We watched them gradually lessening, until they formed a delicate bright cluster, like a swarm of fire-flies on the horizon. On awaking the next morning, we found the stupendous northern precipices of the snowy Olympus full in view, and, landing at the *scala* of Katrin, made our way up to that village. This is one of the stations of the *menzil*, or Turkish post, by which communication is maintained between Salonica and Yanina, and of this means of transit we determined to avail ourselves during this part of our journey, which the *buyurdi* of the Pasha of the former place enabled us to do. In the course of the day a large drove of horses was brought in from the country for us to choose from, and when we had selected the best, we started in the afternoon, in order to break our long ride of the following day. In three hours we reached the *scala* of St. Theodorus, which has been already mentioned as the port of Letochoro, and the place from which the wood of the monastery of St. Dionysius is exported. From thence it is upwards of fifty miles to Larissa, but with *menzil* horses, if your baggage is light, and can be strapped firmly on, so as to admit of trotting, the journey can easily be accomplished in one day. Accordingly, after passing the night at the *scala*, we started early the following day, and skirted the coast at the edge of the Pierian plain, until we reached the castle of Platamona, the ancient Heracleium, which is situated on an elevation overlooking the sea. From its position, and the lofty Turkish tower, which rises, together with a minaret and a single cypress, above the white walls of enclosure, it forms a conspicuous object: the road passes through the valley behind it, by which it is separated from the mountains of the lower Olympus. The name, for which a sufficiently absurd derivation has been suggested—πλατεῖα μονή, or "the level monastery," in allusion to its situation in the plain[1]—is probably taken from the word in ancient Greek (πλαταμών) for a level beach or flat reef of rocks. Between this place and the entrance of Tempe our course lay partly through groves of plane-trees by the sea-side, partly through thickets of prickly palluria, in the direction of the ferry of the Peneius; and when we had crossed that river, we entered the famous vale, and feasted our eyes for the first time on its beetling crags and luxuriant vegetation. The khan of Baba, below Ambelakia, afforded us a halting-place, at the western extremity of the pass.

Once more *en route*, we proceeded for several hours between the lower spurs of Olympus and Ossa, at the foot of which appeared numerous villages, the frequent minarets denoting the religion of these inhabitants of the low-lands. At last the open plain of Thessaly was reached, the monotony of which was only broken by a succession of marshes, caused by the overflow of the Peneius into the lake Nessonis. Through these a rude and slippery paved causeway had been made, such as is usually found in Turkey, in places where the road would be otherwise impassable in winter; but in many parts it was so worn and broken, that we were forced to scramble with difficulty through the quagmire. Frequently we turned to look back on Ossa, which forms the most conspicuous object in this part of Thessaly, for its conical peak is nowhere seen to so great advantage, rising from the mountains which form its base, and descend in grand

---

[1] Meletius, quoted by Leake, 'Northern Greece,' iii. p. 404.

buttresses to the plain. Olympus, on the other hand, presents a far finer appearance from the other side, for here only its southern line of summits is visible, and these are rounded and comparatively shapeless. At length, after passing over a slight elevation, we came suddenly in view of Larissa, lying on the right bank of the Peneius, which is here crossed by a well-built stone bridge, with numerous arches. From without, the place presents an imposing appearance, with its twenty minarets; but within, like most Turkish towns, it is monotonous and dreary. It is said to contain 25,000 inhabitants, and is of importance in other respects from its central position in this valuable province. In consequence of this, it is the head-quarters both of the civil and military governor, and at the time of our visit another Pasha was residing there, who had lately been Derven-aga, or guardian of the passes, in the south of Thessaly. The city contains hardly any remains of classical antiquity.

We put up at the house of a Greek merchant, to whom we had an introduction from Mr. Blunt, the British Consul at Salonica. Here we were soon joined by a numerous company of Greek gentlemen, mostly Ionians, who could speak either French or Italian. They were probably engaged in the silk or cotton trade, for a large quantity of those articles is produced in the neighbourhood, and exported. In the course of the evening one of them undertook to accompany us on a visit to Vassef Pasha, the civil governor, whom we were anxious to consult about the means of transit through the country, as there happened to be unusual difficulty in the matter just at that time. We found him a regular Turkish Pasha of the most approved type, thickset and corpulent in person, dignified and courteous in manner. At first he offered us his own carriage to convey us to Tricala as a favour, which we were afterwards right glad not to have accepted, when we saw the road over which we should have been jolted; on our declining this, he undertook to provide us with horses by the morrow, which might accompany us as far as Yanina. This was good news for us, as there were very few to be met with, in consequence of a large body of troops having just been despatched in the direction of Montenegro, where an outbreak was expected. Not only had all the Government horses been called into requisition, but a clean sweep had been made of all the beasts of burden in the neighbouring districts, to the great injury of the Christian population. After taking leave of Vassef, we proceeded to the residence of the late Dervenaga, from whom we had received great attention and hospitality on a former occasion, when visiting Thaumako, his official head-quarters on the Greek frontier. He was a handsome, agreeable man, but we had learnt in the interval that he was a great scamp, and had been the bane of the district during his term of office, and that it had long been the aim of our representative at Salonica to obtain his dismissal. His last escapade was to carry off three free Greek women, and lodge them in his harem. The house in which we now found him presented a scene of picturesque magnificence, such as is rarely seen in the provinces. The court within which it stood was surrounded by irregular woodwork, and lighted by a flaring torch placed within framework on the top of an upright pole, by the glare of which were seen a number of horses ready saddled, and groups of Albanian retainers lying about in their gay dresses. When we had ascended the stairs and passed through a crowded antechamber, a thick curtain of partition was raised, and we were ushered into the presence of his Excellency, who was seated, in European costume, at the further corner of the divan. He received us with great cordiality, and reproached us for not having taken up our abode with him: and after an interchange of civilities, some of the barefooted attendants, who thronged the lower part of the room, were dispatched for sherbet and chibouques. Evidently the "Robber Chief" (so we had heard him called) was a grandee of the old school, for the sherbet was served in goblets of gilt and elaborately-cut glass, and the mouthpieces of the pipes were surrounded by diamond circlets. In those days, before the reign of cigarettes began, it was customary for the Osmanli of high birth to transfer his jewels from his fingers, not, like the old Roman of Juvenal's days, to his drinking-cups, but to his pipes. He introduced us to his two sons, nice-looking young fellows, the younger of whom he professed to be about to send to England for his education. When we again descended the stairs, after taking an affectionate farewell of our worthy friend, we found three of his magnificent horses waiting at the door, richly caparisoned, to conduct us in state to our night's quarters. Such honours, of course, are not to be had for nothing; for each horse is led by a separate servant, and attended by a separate lantern-bearer, and a liberal *baksheesh* is expected by all.

Early the next morning we received a visit from the Archbishop of Larissa, a fine benevolent-looking old man, with a long grey beard. He had heard that we were on our way to Tricala, and brought us an introduction to the bishop of that place, with whom it was intended that we should pass the following night. When he had smoked the pipe of peace, he rose to depart, and left us free to start once more on our journey. The sun was already high in the heavens, and we had the prospect of nine hours' riding over the parched plain, on which the heat beats down with almost intolerable force at this season of the year. But the traveller who is entertained at private houses must expect to journey in the middle of the day.

Our party was now increased by the presence of four mounted Albanians, whom Vassef had sent to accompany us, as a security against supposed clefts; we would gladly have been rid of them, but we had not the face to send them back to their master. At the end of two hours and a half we reached a flying bridge across the Peneius, by which we crossed to the left bank, and then pursued our course as far as a khan, situated near the low range of hills which forms the boundary between the plain of Larissa and that of Tricala. All through this part, and in fact throughout the whole of Thessaly, the Greek element is predominant; so much so that the decrees and passports of the pashas are regularly written in Greek, since otherwise they would not be understood. After a rest of two hours at this place we were again in the saddle, and the latter part of our ride proved more interesting than the former, as we were continually approaching nearer to the range of Pindus, which lay directly in front of us, and from its fine outline, and the steepness with which it rises from the level plain, formed as grand a mountain barrier as could well be conceived. In one place we saw the peasants threshing the corn by means of horses, which were tied together side by side in a long line, and then driven round and round the threshing-floor. Just at sunset, as we were fording the broad shallow stream of a tributary of the Peneius, the scene for a few minutes was extremely picturesque. Our cavalcade, illumined by the last rays, as they shot forth from behind the dark mass of Pindus, stood out in bold relief, the arms and silver ornaments of our guards glancing brightly in the pure light; at the same time the strange weird rocks of Meteora were in view, standing up in the distance to the north-west. As soon as the last gleam had departed, one of the Albanians imitated the report of a gun, intending thereby to proclaim the arrival of sunset, and the end of the day's fast, and immediately lit his pipe, and commenced smoking greedily. It was an unwonted and perhaps superfluous display of sanctity on his part, for Mahometans by no means rigidly abstain from such things during the Ramazan when on a journey, and for the most part religion sits very lightly on an Albanian.

At last the glittering minarets of Tricala appeared, and we trotted on until nightfall, and reaching the town at nine o'clock, rode straight to the bishop's palace. It was a quaint old structure, faced by a long gallery, on to which, as is usual in such buildings, the various chambers opened; but it was more grotesque in its antiquity than any house that we had seen. The bishop was a fine-looking man of about forty years of age, with an open, smiling, good-tempered countenance; with him we found an old Greek priest, the hegumen of one of the convents of Meteora, who was also a guest. In due course supper was served in the gallery, the carved episcopal chair being placed at the head of the table, while three fine cats seated themselves in silent expectation at the right hand of their master. Like so many Greek ecclesiastics, our host professed liberal opinions, saying that he looked on Christians of whatsoever country as equals in the sight of God, because all believed in the Cross. He also spoke regretfully of the absence of learning from the monasteries on Athos and elsewhere.

Tricala is a large and scattered town without walls, situated at the foot of a hill on which stands a mediæval castle. Its name in ancient times was Tricca, and the change by which it has arrived at its present form is a good example of a process which is found more or less in most languages, but nowhere so conspicuously as in modern Greek. This is the modification of an old name in such a way as to give it a distinct meaning in the spoken tongue. We are accustomed to such changes among the ancient Greeks; thus, to take a familiar instance, the brook Kidron is adapted to the form Cedron, *i.e.* "of the cedars;" but with their descendants it seems almost to have become a trick, for besides the alteration of Tricca into Tricala, "thrice beautiful," and Scupi into Scopia, "the look-out place," which we have already noticed, we find Naxos modified into Axia, "the worthy;" Peparethos into Piperi, "pepper;" Astypalæa into Astropalæa, "old as the stars;" Crissa into Chryso, "the golden," and numerous other examples.[2] Even the Italians, during their occupation of parts of Greece, seem to have been infected with the same mania, as we find them changing Monte Hymetto into Monte Matto, or "the mad mountain;" and Evripo or Egripo, the later form of Euripus, into Negroponte, or "the black bridge," which name was subsequently applied to the whole of Eubœa.

After breakfast the next morning we started for Meteora, which is somewhat less than twenty miles distant. Between Larissa and Tricala the Peneius makes a considerable curve to the south, but beyond the latter city it bends round again in a northerly direction, parallel to the Pindus range, so that at this time we had it on our left hand. In this corner of the Thessalian plain the Cambunian mountains, which bound it on the north, and

[2] *See* Ulrichs' 'Reisen in Griechenland,' i. p. 6.

the Pindus on the west, approach one another at a right angle, being separated by a deep valley, at first about two miles in width, through which the river flows down from the mountains. Just at this angle, on the northern side of the valley, stand the rocks of Meteora, which appeared to us more wonderful the nearer we came to them. From the end of the Cambunian chain two vast masses of rock, with perpendicular sides, are thrust forward into the plain, surmounted at a great elevation by a number of huge isolated columns of various sizes and shapes, on which stand the far-famed "mid-air" monasteries. Between these again, though completely separated from them, another cluster of even larger peaks, wooded round the base, rises at once from the plain to the height of about 800 feet. The stone of which they are composed, and which must have been worn away by some long-continued process of denudation, is a reddish-brown conglomerate, a kind of rock which is singularly well adapted to produce picturesque effect, and is found in some of the most remarkable scenes in Europe,—in the cave of Megaspelæon in the Morea; in the Tajo or chasm at Ronda, in Andalusia; in the extraordinary mountain of Monserrat near Barcelona, and in the Saxon Switzerland. The scenery of the last-named district more nearly resembles that of Meteora than any other in Northern Europe, though the colour of the cliffs is different. At the foot of the central group of aiguilles, on a slight elevation in the midst of mulberry plantations, lies the village of Kalabaka, which we made our head-quarters while we visited the monasteries. This place is on the site of the ancient Æginium, as is shown by an inscription discovered by Colonel Leake; it was a town of some importance from the strength of its position, backed as it was by these inaccessible rocks. Here Julius Cæsar was joined by an important detachment of his forces when descending the valley on his way to Pharsalia. Strange to say, there is no mention of the extraordinary cliffs in any ancient writer, unless indeed they are represented by "Ithome the rocky" (Ἰθώμη κλωμακόεσσα), which is named in Homer in connection with Tricca.

The monasteries which crown the heights above the northernmost of the two rocky tables already described, may be approached from below at one point, where the ground rises for some distance in steep slopes at its foot. Towards this point we bent our steps, skirting the outlying group of peaks behind Kalabaka, and when we had ascended for some time by a winding-path we found ourselves in the midst of thick brushwood, at the base of some of the outermost of the columnar masses. This covert, and the other thickets which are found abundantly in this neighbourhood, are reputed to be a great resort of robbers; (*forty* thieves they were said to be, for the old Oriental round number is still regularly in use), and as a protection against these a number of Turkish guards had been sent to accompany us. The track became still steeper as we continued to ascend among the rocks, which assumed new forms at every turn, while between them appeared glorious views of the plain, the river, and the distant mountains. At length we reached a small platform of ground beneath the mass of conglomerate on which the Great Monastery of Meteora is placed. The sides of this were everywhere precipitous, and the ascent was to be accomplished by means of a net and a rope, the end of which hung suspended from a projecting loft right over our heads. We shouted, in expectation of the same ready reception to which we had been so accustomed on Athos, especially as we had brought with us a letter of introduction from the Bishop of Tricala ; but, to our surprise, no reply was made. A pistol was then fired off, by way of attracting attention, but, though it re-echoed among the surrounding cliffs, it failed to elicit any response from above. Our dragoman then commenced an appeal *ad misericordiam* : " Holy fathers, we don't want your wine, we don't want your bread, we don't want anything of yours, we only want to see your monastery, and we have a letter from the bishop." Still all was silence ; but, shortly after, one of our attendants saw some of the aërial occupants peering through the openings of the loft, and immediately renewed the appeal —" O, holy fathers ; O, father Stephen, these are respectable men, they are English milords, they are distinguished personages, they are of royal extraction "—a cumulation of honours at which we could not help laughing, and we thought of Mr. Curzon's description of himself as first cousin of the Emperor of the Franks. Our position was ludicrous enough, for the monks felt that they were masters of the situation, which undoubtedly they were ; but it was excessively provoking, and all the more so when we found that the same panic had seized the monks on the neighbouring height of Barlaam, and that though we brought an introduction to them from their own hegumen, they would pay no attention to us at all. There was nothing for it but to try our fortune elsewhere, so we rode along the heights through scenery of the utmost grandeur, in the direction of the convent of St. Stephen, which occupies one of the peaks of the other group.

The position of this monastery is different from that of any of the others. The isolated peak on which it is built does not rise higher than the level of the ground behind, but is separated from it by a narrow chasm, some

60 feet deep, over which a drawbridge is thrown, and forms the only means of access. It overhangs the plain at the height of a thousand feet, and the views which it commands are indescribably magnificent, comprehending the weird forms of the outlying rocks, the Peneius, the chain of Pindus, whose stupendous buttresses are here seen in profile, and the wide expanse of the plain of Thessaly, bounded at last by the heights of Othrys, which are sixty miles distant. No one who, in descending on the southern side of the Alps, has seen beneath him—

> "—— outspread, like a green sea,
> The waveless plain of Lombardy,"—

will easily forget the effect of that extraordinary expanse. But this level is in some respects even more remarkable. For while the Italian plain is intersected by dikes and hedges, which, however minute they may appear at so great a distance, have in some degree the effect of breaking up the uniformity of surface, here there is not a single object to interfere with the one unbroken level, which in spring-time is for the most part clothed with waving corn, and in summer presents an undivided area of dry and yellow soil.

As we approached the monastery, having suddenly emerged from the neighbouring thickets, two monks, who were standing outside the drawbridge, on seeing an armed company approaching, retired within the building and closed the gate. This looked ominous, but when we rode on in front of our party, to show that our intentions were peaceful, the hegumen came out and invited us to enter. The drawbridge was a rickety concern, only about three feet wide, without any balustrade, and with wide interstices between the cross-planks. Nicola, our dragoman, who, though a plucky fellow, had a bad head, required to be conducted across by two monks. Still worse was the plight of one of the guards. He extended his arms like a rope-dancer, stepping delicately six inches at a time, and when he felt it trembling beneath him, his "cream-faced" look of open-mouthed horror was not easily to be forgotten.

The hegumen, a brisk little man with crisp brown hair, ushered us into a delightfully clean apartment, and made us as comfortable as possible. As for our Turkish attendants, we were determined that they should not pass the night in the convent, so we sent them away to Kalabaka, with orders to return in the morning. The monks appeared to live in continual dread of the robbers in the neighbourhood. In the evening, when we went out to a point a few hundred yards outside the walls to make a sketch, they strongly remonstrated with us on our imprudence, saying that there was every chance of our being carried off. This was the case with numbers of their own body, when they were forced to go to Kalabaka for provisions, and they were only released on payment of a heavy ransom. Even the dwellers in the plain seemed not to be safe from their incursions. At the time of our visit a gentleman of Kalabaka was residing in the monastery with his daughter and others of his family, whom he regularly brought there during the summer months for fear of these marauders, while in the winter, when the cold forced them to disperse, he returned to his estate. In the year 1831 a number of them stormed the Great Monastery of Meteora, bound the monks, and plundered the convent. How they got up there it would be hard to say, but it is equally difficult to answer the question, how the original inhabitants scaled those rocky columns, and how the materials were carried up of which the buildings are composed. Of the existence of these gentry we had satisfactory evidence after nightfall. As we were sitting at supper with the hegumen, we heard a loud shouting and yelling outside the walls, and inquired what it meant. "Oh! it's of no consequence," he replied; "it's only some of the clefts, who want us to hand them out some provisions; but it's all right, for the drawbridge is up, and they are on the other side, so let us go on with our supper." We then discovered the advantage of the position of the convents of Meteora.

From the hegumen we learnt that, while there were originally fourteen monasteries, there are now but seven existing, and only four of these are inhabited by more than two or three monks. These are St. Stephen, Holy Trinity, Barlaam, and the Great Monastery of Meteora. The convent of St. Stephen was founded by the Emperor Cantacuzene in the middle of the fourteenth century, and contains fifteen monks, five of whom are priests. As regards their constitution they are entirely independent, owing no allegiance either to the Patriarch of Constantinople or to the neighbouring bishops. Their decline is owing to their poverty; thus this society at the time of our visit only possessed a few farms in the adjacent parts of Thessaly, and some in Wallachia, which now they have probably lost. To the same cause they attributed their ignorance, for though they expressed a desire for learning, they had no means of introducing or fostering it. Their rules are not nearly as strict as those observed on Athos; they are not forbidden to eat meat, and, as regards the services, prayers are said in the church three times a day at ordinary seasons, and in Lent seven times. In winter the cold is very great, so that the monastery is often covered with snow. There is said to be a great abundance of game in the neighbourhood, and boars, stags, and wolves.

Before retiring to rest, as it was a brilliant night, I got one of the monks to conduct me to a point at the extremity of the rock on which the monastery is built, from which I could get a view of the valley; and the scene which I saw was truly sublime. The cliffs here descend a thousand feet in an unbroken fall to the plain; far below me appeared the dim forms of grotesque rocks, and beyond this, the expanse of the valley, with two silver streams of the Peneius lighted up by the moon, and the broken outline of Pindus, opposite which on one side loomed out a group of the huge black columnar masses of Meteora. What a spot, I thought, for religious contemplation, if such exists among the monks!

After a good night's rest on the cleanest of coverlets, we proceeded to visit the buildings, which cover nearly the whole of the detached rock. Owing to the narrowness of the space they are packed away with great irregularity, just where convenience suggests, and the imposing order which is observed in the arrangement of the convents of Athos is here wanting. The principal church,—which is, as usual, in the Byzantine style,—corresponds in shape to those found in the larger Greek monasteries, but in other respects it forms a singular contrast to them, being almost destitute of ornament, except in the altar-screen, which is very handsomely carved in wood with figures of birds and flowers. There is also an older church, covered inside with frescoes of saints, many of which are disfigured by marks of the swords of the Turks. The only relic is the head of St. Caralampus, a saint who was martyred near Ephesus: the casket in which this is kept is covered with filigree silver-work, and in shape somewhat resembles a crown surmounted by a cross; it is a rarely beautiful work of art. The old refectory is worthy of notice from its small dimensions and vaulted roof. We also visited one of the cells, which was scrupulously clean; indeed cleanliness was the distinguishing feature of this monastery. There are several cisterns within the walls.

We had sent on a messenger to the convent of Barlaam, to present our letter, and request admittance. The hegumen had now returned, and sent back apologies for the treatment we had received on the previous day, saying that there were only two boys within at the time when we passed, and they could not admit us without orders. We were glad, at all events, to know that we should at last be able to ascend to one of these eagle's nests. Meanwhile our guards had made their appearance, and in their company we retraced our steps along the heights, until we reached the foot of the rock on which the monastery stands. The face of this is absolutely perpendicular—an epithet, which though often applied to cliffs, is seldom literally true, as it is in this instance. Looking up we could only see the end of the rope, by which we were to ascend, hanging from a block, with a great iron hook attached to it; but shortly after this was let down, accompanied by a strong and capacious net. The net was unhooked by our attendants and spread out upon the ground. I immediately seated myself in the middle of it, with my legs crossed under me *à la Turque;* but the monks called out that two should get in together, so one of my companions joined me, and when we had been placed *vis-à-vis*, and our legs intertwined in a marvellous manner, the net's meshy folds were collected around us and the hook passed through them; a shout from below, a pull from above, and we found ourselves swinging in mid-air. There was a twist on the rope, and consequently we went round and round at first, like a joint of meat suspended from a bottle-jack; and we found it highly expedient to cling on tightly to the sides of the net, so as to save ourselves from tumbling into a confused heap in the bottom. Up, up we went, with an easy and gentle motion, and as we looked down between our legs the rocks, and trees, and deep gorges, appeared to be receding below us; until, at the end of three minutes, we were wound up to the block, and the good fathers, putting out a hooked pole in default of a crane, fished us in, turning us over in the process, and laid us in a helpless condition on the floor. We were then disentangled and helped to our legs. With grim gravity the monks addressed us with the customary salutation of "Welcome to the end of your journey" (καλῶς ὡρίσατε), to which, when we had shaken ourselves into shape, we equally gravely replied, "We are happy to have reached you" (καλῶς σας εὑρήκαμεν). When we looked down the precipice the depth was appalling; the monks called it 222 feet. The rope was worked by a windlass, and on examination we found it was frayed in a very unpleasant manner.

Besides the net there is another way of ascending to the monastery, which is ordinarily used by the monks. In one part of the rock, where the cliffs are not quite so perpendicular, wooden ladders are attached to its face, conducting to the entrance of a tunnel passage, which leads into the interior of the monastery. One length of the ladder hangs loose, and can be hauled up, so as to cut off communication with the world below. One of our party ascended by this way, and experienced no unpleasant sensation, excepting a slightly uncomfortable vibration of the suspended joints. He was followed by Nicola, who could not make up his mind to the net, and in reality chose the more difficult process. When he had reached a narrow ledge of rock, at a considerable elevation, he stopped to take off his boots, which were slippery; and for some time we were frightened about him, knowing that he had not a good head: soon, however, he was able to start again, and, with some assistance from the monks, reached the mouth of the tunnel in safety. Next came the turn of the guards, two of whom summoned up courage to mount by the net, and were swung up in the midst of the jeers and cheers of their companions, and promiscuous discharges of firearms. It was a curious sight to see this bundle of red and white baggage (for this was the appearance their Albanian dresses assumed when huddled together in the net) gradually ascending from below.

We then proceeded to lionize the monastery. It takes its name from a hermit called Barlaam, who fixed his abode on this rock, and has since been enrolled among the saints of the Greek calendar. Dr. Wordsworth supposes[3] that it was called after the Calabrian monk of that name, who was so prominent in the 14th century as an opponent of the doctrine of the divine light of Tabor; and if this were true we should have the curious spectacle of two rival convents on two of the heights, representing the two sides of that controversy, for that of St. Stephen, as I have mentioned, was built by the Emperor Cantacuzene, who was one of the strongest supporters of the doctrine. But it is singularly unlikely that one who was declared a heretic, should be commemorated in a monastery of the Orthodox Church; and the account given above, which is derived from Colonel Leake,[4] is far more trustworthy. In Russia the name Barlaam is

[3] 'Greece, Pictorial and Descriptive,' p. 208.
[4] 'Northern Greece,' iv. p. 541.

not uncommon. The principal church is a handsome building, and contains some of the finest Byzantine frescoes I have ever seen. One, of the 'Repose of the Virgin,' is especially beautiful, and the grouping of the figures is truly artistic, a thing very rarely found in works of this school. There is also an older church; a refectory, near the entrance of which is a buttery-hatch on one side; a kitchen, resembling on a small scale some of those on Athos; and a library, in which the books were carefully arranged. Within a small ante-chamber to the library the sacred relics were kept, all being enclosed in finely-wrought silver caskets; among them was shown a hand of St. Chrysostom—a circumstance which aroused Nicola's scepticism, as he had just seen his other hand on Athos, and did not much believe in two hands of the same saint being kept in different places. But the most original building was the hospital,—a small square structure, with stone benches running all round it, and in the middle a hearth, over which rises a lofty chimney supported by four columns. There are also cellars of considerable size; and in one part, outside the walls, there is a rough garden for vegetables hanging about the rocks. The passages by which it is approached are entered by remarkably low doors, and the establishment in every way gives an idea of great strength and powers of defence. It is now tenanted by a dozen monks.

When we had partaken of the customary refreshments, and presented a donation to the monastery, we prepared to descend again. This was the most exciting part of the whole proceeding, and bore a strong resemblance to going to execution. There stood the monks, our executioners; there hung the rope from which we were to be suspended; there, above all, was the dreadful abyss over which we were to swing. However, short work was made of us. The net was spread at the edge of the loft; two of us were placed in it as before, the meshes were gathered round us, and the hook passed through them; and then off!—we were bundled out, and in a minute and a half had safely reached the ground. Then came our other companion with Nicola, who clawed hold of him in a convulsive manner, and gave vent to hysteric chuckles during the descent; and, last of all, the bundle of red and white clothes, in the shape of the two Albanian guards. In the meanwhile we had despatched a monk to the Great Monastery of Meteora to request admittance, but they still refused; and we discovered from the hegumen of Barlaam that that convent was in a state of great confusion, the caloyers having risen in insurrection and expelled their chief, and being consequently suspicious of any strangers. So we mounted our horses, and in half an hour had returned to Kalabaka.

The curious mode of ascent and descent which has just been described seems not to have been altogether uncommon at one period, and to have been in use in castles as well as monasteries. In the ballad of the 'Beauty of the Tower,' already referred to, the mention of it is introduced. There the leader of the band of Turks who come to storm the castle, having disguised himself in the dress of a monk, presents himself before the entrance; on which the following dialogue ensues:—

"*The pretended monk.*—Open, open, thou door: door of the beauteous
    lady;
Door of the dark-eyed lady, the princess.
The lady.—Thou art a wicked Turk, a Koniarate;
    Away, or they shall kill thee; away, or they shall hang thee!

Monk.—No, by the cross, my lady! no, by the Holy Virgin!
   I am no wicked Turk, I am no Koniarate;
   I am a holy caloyer from a monastery.
   I am dying of hunger; have compassion on me.
Lady.—There, give him the loaf. Now, go about your business.
Monk.—Nay, lady, suffer me to visit the church, and offer a prayer.
   Open, open, thou door: door of the beauteous lady;
   Door of the dark-eyed lady, the princess.
Lady.—Let down the hook, then, and take him up.
Monk.—O no: my serge is rotten, and will tear.
Lady.—Let down the sack, then, and take him up.
Monk.—O no, good lady, not the sack; I shall turn giddy."

Then follows the *dénouement*. The gate is half opened. The pretended monk prevents it from being closed again. His company rise from their ambush, and the castle is entered and stormed.

As to the monastery of Barlaam, which we have been visiting, there is an amusing cleftic ballad, which relates an incident that took place there, resulting in the abduction of the hegumen. It illustrates the ill-feeling which usually existed between the clefts and the monks, though occasionally, as at Sparmos on Mount Olympus, we find them in league with one another. The hero of it, Eutimios,[5] or, as he was popularly called, Papa Thymios—from his having been a priest in early life—was the son of a captain of Armatoles in the mountains near Tricala, and at his death succeeded to his father's office. This was during the lifetime of Ali Pasha, and for a time he maintained the same good understanding with that chieftain that his father had done; but when Ali began to put down the Armatoles in other parts of Thessaly, he revolted, and began to plunder and murder Turks in the neighbourhood of his district. At this time

---

[5] *See* Leake, 'Northern Greece,' iv. p. 542.

he laid the monasteries of Meteora under contributions, and in consequence of their having supplied him with provisions, though against their will, several monks from each monastery were afterwards imprisoned by Ali at Yanina, and had, no doubt, to be ransomed at a heavy sum. He was ultimately forced to take refuge in the islands, and having been captured and taken to Constantinople, was given up to Ali and executed by him after a variety of tortures administered at intervals. The ballad runs as follows:—"

> "Ye singing birds of Grevenó, and nightingales of Metzovo,
> Ye know the Father Thymios, the holy priest, ye know him:
> When he was young he learnt to read his letters like a scholar:
> Now that he's old, he's shown himself a noble robber-captain.
> In every fort he set his foot, in every monastery,
> Except the convent of Barlaám; there he could never enter.
> For it was perched upon the cliffs, high on a rocky column.
> He called to the Hegumenos, and once again he called him.
> 'Come down, my Lord Hegumenos, we want you as confessor:
> One of our company is sick, his death is fast approaching.'
> The abbot took a taper-light, and in his stole arrayed him;
> Then he descended to the court, that so he might confess him.
> 'Long life to you, Hegumenos.' 'Welcome, brave Palikari.'
> 'Take off your gown of serge, good man, and of your stole divest you;
> 'Tis heavy work to walk in them: you'll have to pass through thickets.'
> They bound his hands behind his back, and led him from the courtyard;
> Four of them went in front of him, and four of them behind him."

" Passow, 'Pop. Carm.' No. cx.

CHAPTER XXV.

PINDUS.

Upper Valley of the Peneius — Malacassi — Defeat of the Greek Insurgents in 1854 — Pass of Zygos — Fountain-head of Rivers and Meeting-point of Races — Metzovo — The Wallachs — Origin of the Name — Evidences of their Roman Descent — Customs and Beliefs — Roman Colonies in Dacia — Migration south of the Danube — Subsequent History — Bulgaro-Wallachian Kingdom — The Wallachs of Thessaly — Present Condition.

IN the afternoon of the same day we left Kalabaka, and proceeded in a north-westerly direction up the valley of the Peneius. When we reached the river, we found it here a narrow stream flowing with a swift current over a broad stony bed, and so encumbered with floating timber from the cuttings above, as to require more than ordinary care at the ford. We continued to skirt its banks during the whole of the evening. The level plain, which runs in for some distance, at length merges in a narrow vale, the scenery of which is extremely beautiful, from the height of the surrounding mountains and the luxuriant growth of the oaks and plane trees. At intervals huts of branches appeared beside the track, which proved to be stations for guards; and as we trotted along, two slim figures slipped out from each of these, and ran along near us, until relieved at the next post. They were spare, sharp-featured fellows, in dirty fustanellas and white skull-caps, and armed with long guns—picturesque characters, but suspicious in appearance, had we not been assured of the duty they perform. Sometime after nightfall we reached the khan of Malacassi, one of a number of similar buildings along the road from Tricala to Yanina, affording nothing but shelter, bread, and arrack, but superior in their accommodation to those of northern Albania. It was situated in a wild spot, where the river rushes by far below, brawling over the rocks in a succession of small cataracts. The single chamber, which was built above the stable, was black from the smoke of pinewood fires, and had a gallery open to the air, where we slept refreshingly. It was the same apartment in which Mr. Curzon describes his robber-guard as having taken up their quarters, when he was returning from his expedition to the Meteora. The village of Malacassi lies away at a little distance to the north, and is occupied by Wallachs, of which race a large part of the population of this part of the Pindus is composed, the settlement dating from the 11th century.

We passed a quiet night at the khan, nor were our thoughts disturbed by any scenes of fighting, unless it was the campaign of Julius Cæsar, who must have passed this spot when on his way to engage in his last great battle. But had we been able to forecast the events of the following year, we should have seen this lovely valley full of confusion and bloodshed. It was now the 16th of June, 1853. On the same day of the same month in 1854 the khan was a ruin in the midst of blackened timbers, having been destroyed by the Greek insurrectionists, who had taken the opportunity of the first summer of the Russian war to cross the Turkish frontier, and overrun the whole of Thessaly. On the evening of that day it formed the *rendezvous* of the forces of the Turkish commanders, Fuad Effendi and Abdi Pasha, the former of whom had hitherto been

occupied in quelling the rising in Epirus, from whence
he had just crossed by the pass of Zygos, while the
latter had arrived, in accordance with a preconcerted
plan, from the district of Grevena to the north, in the
upper valley of the Vistritza (Haliacmon), where an important engagement had lately taken place. On the
following day they marched down the valley with their
combined forces, consisting of 3000 regular troops with
six guns, and 4000 irregulars, in the direction of Kalabaka, where the insurgent bands were stationed. Shortly
before this, the Greeks had gained the only success
which fell to them during the campaign, having routed
a force of Arabs under Selim Pasha, who seems to have
penetrated imprudently into the valley, and to have
allowed the communication with Tricala to be cut off in
his rear. Ultimately he was obliged to force his way
through the enemy, leaving all his baggage and three
guns behind him, and losing several hundred men in his
retreat, numbers of whom were drowned in the river.
The insurgents were greatly elated by this success, and
determined to await the important force which was
coming against them. Kalabaka was the scene of the
final conflict. One would wish at that moment to have
been an inmate of one of the monasteries of Meteora.
What a splendid spectacle must have presented itself to
one looking down from those mountain eyries! The
whole battle-field lay beneath, a thousand feet below
them, and every move of the combatants must have been
traceable with the utmost distinctness. And then, the
strain of expectation, the swaying to and fro between
hope and fear, by which their minds must have been
agitated while they watched the struggle! Did they see
a cross of light appear in the heavens, as had occurred

on Athos on a similar occasion? Or did they behold, as Pouqueville[1] relates that their predecessors in these very monasteries beheld, a horseman in bright raiment and glittering armour driving the infidels before him over the plains of Thessaly, and vanishing at length amid the ruins of a church of St. Michael, whence sounded forth the war cry of the heavenly host, "Hosanna in the Highest"? It is probable enough that such portents were seen; but, if so, the visionaries were doomed to disappointment, for they were to behold the discomfiture of those in whom their hopes were centred.

To return, however, to the forces in the plain. The Greeks had occupied two positions: the one was at Kalabaka itself, backed by the great rocks, a post which they had carefully fortified, and on which were seen the tents they had taken from Selim Pasha; the other was a point on the lower heights of Pindus, where a projection, running out at an angle to the main chain, intervenes between Kalabaka and Tricala. The Turkish forces had encamped for the night on the right bank of the Peneius, and in the morning, at a council of war held under a plane-tree in the dry river bed, it was determined to make a demonstration against Kalabaka with the regular troops, and send the irregulars to attack the outlying position. Among these latter the Albanian Ghegs were the most conspicuous; and when they reached the spot, they carried the hill nearest the river by assault under a murderous fire, and then with the assistance of the rest of their party, cleared the other heights, sending the enemy flying in all directions. As might be expected in a force of half-trained insurrectionists, the effect of this first engagement, which could be seen from Kalabaka, was to

---

[1] Quoted in Bowen's 'Mount Athos, Thessaly, and Epirus,' p. 160.

strike panic into the main body; and when the regulars advanced, after fording the river, they succeeded in storming the Greek fortifications. The defenders retired within the gorges between the great cliffs, and there maintained themselves for some time, but at last they were expelled from these also.[2] The blow they received on this occasion was fatal to their cause, and from that time the Greek insurrection in Thessaly was virtually at an end.

About sunrise the next morning we left the khan, on our way to the pass of Zygos ("the yoke") the Lacmon of antiquity, by which the central range of the Pindus is here crossed. It took us about two hours to reach the summit of the ridge, during which time we mounted by an easy and gradual ascent, looking down into two deep valleys below us, sometimes towards the Peneius on our left, sometimes on our right towards its tributary, the stream of Malacassi. In some parts of the way large forests of beech and pine are passed, and from the upper regions the views are extensive over the plains and mountains of Thessaly, the prospect on that side being closed by the great Olympus. The height of the pass is estimated by Boué[3] at 5063 feet, and even at this season snow was lying on the ridge in some few places. The summit of the *col* is striking from its sharpness, and from the rocks which rise on either side of the track, thus forming a sort of gateway, through which you pass into Albania. In other and more important respects also—both as a fountain-head of great rivers, and as a point of separation of various races, these mountains of Zygos are a noteworthy position, so that more than one traveller has spoken of them as among the most remark-

---

[2] *See* the account of an eye-witness, given in the 'Times' for July 11, 1854.   [3] 'Recueil d'Itinéraires,' ii. p. 60.

able localities in the geography of Continental Greece. When Virgil represents the shepherd Aristæus as following up the stream of the Peneius to its source, and there finding the home of his mother, the river goddess, he describes him as seeing the head-waters of all the great rivers on the earth, flowing beneath the ground at this spot. He could not have fixed on a place more suitable to embody such a conception, as it is the birthplace of the five most important rivers of Greece. Besides the Peneius (Salampria)—which we, like Aristæus, have traced upwards from Tempe—the Achelous (Aspropotamo), which flows southwards to the neighbourhood of the Corinthian gulf; the Arachthus (Arta), which finds its way into the Ambracian gulf; the Aous (Viosa), which we have seen flowing under the walls of Tepelen, on its way to the Adriatic; and the Haliacmon (Vistritza), which pours its waters into the gulf of Salonica; all have their origin in these mountains. Nor is it less remarkable in an ethnographic point of view, from the numerous races which here approach one another. For whilst in the plains of Thessaly to the south-east we have left behind us a predominant Greek population, and Albanians are found in the mountains to the north-west, the north-eastern slopes are tenanted by Bulgarians; so that these three nationalities would meet at this point, were it not that a fourth is interposed between them, in the shape of the large Wallachian colony, which I have already spoken of as occupying these mountains. If we may compare small things with great, it holds in European Turkey the same position which in Asia is held by Mount Ararat, that great boundary-stone between three races and three empires—the Russian, the Turkish, and the Persian.

The western side of the pass is strikingly different

from the other, and its aspect is thoroughly Albanian, from the wildness and confusion of its mountain masses. Here, too, the descent is very steep, and as you look down the deep valley below you, and trace the windings of the path along the sides of the mountains, you are forcibly reminded of some of the inferior passes in Switzerland. Upon our right appeared the town of Metzovo, built in terraces upon the hill side. We did not visit it, as it lay out of our route, but halted at a khan on the opposite side of the Arta, which flows beneath it. It is a large place, and its population, which is entirely Wallachian, amounts to not less than 5000 souls. We have already met with members of this race in some of the great Turkish towns, such as Monastir and Calcandele, to which might be added Salonica, Larissa, and many others. We have also noticed them, either in a nomad or settled state, in various parts of the country—on the Scardus pass above Prisrend, on Mount Olympus, and at Vlako-Livadi on the western side of that mountain, and in the neighbourhood of Berat in Albania. They are also found at other places in the neighbourhood of the Pindus chain, in the valleys of the Balkan and other districts of Thrace and Macedonia, and even as far south as the Morea. But no point that they occupy can be regarded as so central, or so nearly representing their head-quarters, as that where we have now halted; so here it may be worth while to give a brief account of their origin and characteristics. As they are the kinsmen of the Wallachians,[4] who inhabit the Danubian Principalities, these also will be included in any remarks we shall make on the race in general,

[4] The term Wallach, Vlach, or Vlakiote, is commonly applied by English writers to the tribes in Thessaly, Macedonia, &c., while their brethren of the Principality are usually called Wallachians.

though our primary object is to describe the condition and relate the history of those south of the Danube.

The name of Wallach is not acknowledged by the people themselves, but, like the ordinary designation of many other races, was imposed on them by foreigners. At what time, and by whom, it was first given there is no evidence to show; but it is found at an early period, and it is tolerably clear that it must have been given by Teutonic or Slavonic tribes, probably the latter. The word, whatever its derivation,[5] is found, with slight variations in form, as the title by which the Teutonic races called the peoples of a different stock, on whom they bordered, especially those of Celtic or Roman origin; and in this way it came in course of time to signify simply "a foreigner." Thus among the Anglo-Saxons Welsh was the name given to the Cymry; and amongst the Germans France was known as Walho-land, and throughout a great part of the Middle Ages Wälsch was their name for an Italian, and is so to some extent even to the present day. In this way the French-speaking population of the Netherlands were called Walloons; and the Swiss canton, which contains the Rhone valley, got the name of the Canton Wallis, as being inhabited by Italian foreigners—a name which has subsequently been corrupted into Vallais. And similarly the inhabitants of those valleys of the Grisons where the Romansch dialect is spoken are called Churwelschen. The word seems to have passed from the German to the Slavonic tribes, and by them to have been applied to this eastern race, with whom they were brought into contact. Among themselves all the various branches of the Wallachian

---

[5] According to Schafarik ('Slawische Alterthümer,' i. pp. 236, *seq.*) it was originally the same word as *Gael, Celt*, &c., and at first was only applied to tribes of Celtic origin.

family are called Rumuni, or Romans, with the single exception of this colony in the Pindus, among whom we have now halted: they style themselves Armeng, a name which is difficult to interpret, though it may possibly be a corruption of the same word. Those that dwell south of the Danube are frequently known by the name of Tsintsar, an appellation which forcibly recals the original "shibboleth," being derived from "tsints," which is their pronunciation of "chinch" (five, *quinque*). It is, of course, a name of ridicule, and unwelcome to their ears, as is also the case with the title Kutzo-Vlachs, or "lame, halting Wallachs," which refers, I believe, to the same, or a similar, defect of speech. The Albanians call them Tjuban (which is the Turkish word for "shepherd") on account of the pastoral habits of a great part of the tribe; they also use the name Agoghi,[6] which is probably applied only to those who exercise the trade of carriers.

The name of Rumuni, which this people apply to themselves, though it at once indicates a Roman connection, does not necessarily prove that they were descended from the Romans. The same title was assumed by other peoples who were admitted to the Roman franchise—a fact of which we are sufficiently reminded by their neighbours, the modern Greeks, calling themselves by this name ('Ρωμαῖοι). Indeed, it would be hard to find a stronger proof of the value that at one time attached to the expression "*civis Romanus sum*" than the existence at the present day within the Turkish empire of two races still calling themselves by the name of Roman.[7] But if we want more distinct proofs that the

---

[6] See above, on Prisrend, vol. i. p. 341.

[7] In the Eastern Empire, when the name 'Ρωμαῖος was first used, it seems to have been adopted in order to express a further idea, viz., that of "Christian," as opposed to "Ελλην, which had come to bear the meaning of "Pagan."—See Heilmaier, 'Entstehung der Romaischen Sprache,' p. 9.

Wallachs are akin to the Romans, we have it in their language, which is certainly derived from Latin. On this subject it may suffice to say, that in the opinion of good critics it is older, that is, more primitive in its structure, than Italian, French, or any other of the cognate languages, except only the Romansch of the Grisons. It is also found that the Latin element in the language is preserved in the purest and most unchanged condition among the lower classes in remote localities; a clear sign that it has descended from early times, and not been impressed at a later period. Another conclusive proof exists in the numerous customs and beliefs common to them and the Romans. Of these a considerable collection has been made in the Banat in Hungary, where a large Wallachian population has long been settled. It may be worth while to introduce some of them here.

Amongst the Romans the "strix" or screech owl was supposed to suck the blood of young children: the Wallachians apply the name Strigoi to a class of evil spirits, which are supposed to bear ill-will to infants. When a child is born, those who are present are accustomed to cast a stone behind them, saying at the same time, "This in the mouths of the Strigoi."

The plant which we call woodruff (*asperula odorata*) was known by the Romans as "herba matris sylvæ." "Muma padura," or "the mother of the woods"—a title by which the Wallachians designate a beneficent forest-spirit, who is especially the protector of children—is also their name for this flower.

Thursday and Friday are still in a way consecrated to Jupiter and Venus, the divinities to whom they were dedicated among the Romans. During a part of the spring the former day is regularly kept holy, in honour of the god of thunder, that he may not send hail and stormy weather. Friday, on the other hand, is especially observed by women, and is regarded by many of them with greater reverence than Sunday. On that day they avoid working with a sharp or pointed instrument, such as a pair of scissors or a needle.

Amongst the Romans the Lares, or household gods, were regarded as the departed representatives of the family, and were believed to preside over the house and hearth, in which character offerings were regularly made to them at meals. Similarly among the Wallachians every family is dedicated to a saint, whose festival day is observed, even by the poorest, with the greatest care. The whole house is then cleaned, the furniture washed, and a feast prepared, at which especial mention is made of dead relations. They are invited to table with earnest prayers and invocations, and places are left for them, with covers, on which wine, salt, and bread are laid.

Other ceremonies connected with the dead betray a no less striking resemblance. A small piece of money is placed in the hand of the corpse before burial, as a guarantee of the reception of the dead below, in the same manner as the Romans placed it in the mouth. The Roman custom of mourning for the departed with the head uncovered is still commonly observed. One old man, whose son died at twelve years of age, is known to have gone bare-headed for the rest of his life, in consequence of a vow to this effect. It is also customary, on the anniversary of a burial, to make an offering of bread and wine on the grave, the wine being poured over it.

Other points of similarity had been observed as early as the fifteenth century by Chalcocondylas, the Byzantine historian, who remarks that the Wallachs—that is, those living south of the Danube—not only spoke a language like that of the Romans, but also bore a singular resemblance to them in their habits, mode of life, arms, and household implements.⁸ We also find that in the Middle Ages the people themselves had a consciousness of some original connection with Rome, which was even turned to some political account. Thus, Basil, Archbishop of Zagora, writing to Pope Innocent III., in the year 1204, reminds him that the Wallachs in Thrace were of Roman blood; and the same Pope, when negotiating with King John, one of the early sovereigns of the Bulgaro-Wallachian kingdom, pays him the compliment of saying that he and his people drew their origin from Rome.⁹

Let us now endeavour to trace their history, as far as it can be discovered. The extensive plains to the north of the Danube, which form the Principality of Wallachia, together with considerable portions of the neighbouring countries, bore in Roman times the name of Dacia, and were not reduced into the condition of a Roman province until the reign of Trajan, in the beginning of the second century Anno Domini. In consequence of the wildness of the tribes by which it was inhabited, the depopulation which the war had caused, and the importance of its position as a bulwark against the barbarians, who were already beginning to threaten the frontiers of the empire,

---

⁸ Chalcocondylas, quoted by Thunmann in his 'Untersuchungen über die Geschichte der östlichen europäischen Völker,' p. 345. It is remarkable, as showing how little our knowledge of the races in the interior of Turkey has advanced within a century, that this book, which was published in 1774, is still in most respects our best authority on the Wallachs south of the Danube. His information with regard to their dialect was derived from a book printed at Venice by a person of that race, who lived at Moschopolis, a large Wallach village not far from Ochrida. This place seems to have produced a number of men of some learning at that period. About the people generally there is excellent information in the introduction to Schott's 'Walachische Märchen,' and in the chapter on their superstitions, from which are derived the notices of their customs which I have given above. ⁹ Ibid., p. 345.

the Roman system of planting colonies in conquered countries was here carried out even more extensively than usual. Eutropius tells us that Trajan transplanted thither innumerable bodies of men from the whole Roman world. That these settlers outnumbered the native population is highly improbable; that the original inhabitants were driven out or exterminated, as some have supposed, is almost incredible; for all modern historical research goes to show that it is next to impossible entirely to depopulate a country which is inhabited by a people above the condition of savages. But as the Roman domination only lasted until the time of Aurelian—that is to say, during a period of about 160 years, we have here a remarkable proof of the claim that Latin has to be classed among what Mr. Hallam calls the "aggressive languages;" in other words, of the power it possessed, from the completeness of its syntax, the richness of its vocabulary, its historical associations and embodiment in literature, to overpower and supplant any weaker language that came in its way. The result of the process is the Wallachian tongue as we find it now: this at least is the only way in which we can explain the existence of a language so manifestly the daughter of Latin in this remote region.

It would be an interesting task to trace the later history of this country, after its evacuation by the Romans, and that of the neighbouring lands which were occupied by this population. But this would be beyond our present purpose. We must now turn our thoughts to the regions south of the Danube, with which we are more immediately concerned.

At the time when the Roman forces finally abandoned Dacia, a large number of the colonists who had been established there, retired south of the Danube, and dwelt in the country between that river and the Balkan. As their migration took place in Aurelian's reign, this district received the name of the Dacia of Aurelian, in contradistinction to the former province, which was called the Dacia of Trajan. At this point, however, they pass out of sight, and are lost to view in the midst of the flood of invasions by Huns, Avars, and numberless other barbarian tribes that overran the countries that border on the Danube. "The historian," Thunmann complains, "is often as unjust as ordinary men, in despising the unfortunate." He could not have found a more appropriate sentiment with which to introduce an account of the southern Wallachs. Though there can be little doubt that they were the descendants of these colonists (for, besides other proofs, the uniformity of the Wallach language wherever it is found, whether in Hungary, Moldavia, Thrace, or Thessaly, shows that it was impressed on the whole race at one time), yet their name does not occur until the eleventh century; and after that time they are but sparingly mentioned by writers, whose pages are filled with minute details, often extremely uninteresting, of the actions of emperors and usurpers. To this it must be added, as an additional cause of their disappearance from history, that they differ from all the other races now inhabiting European Turkey, in the want of strong national, as distinguished from tribal, feeling. Though their communities have remained unmingled with the other nationalities, yet they never have shown any disposition for united action amongst themselves, and have been comparatively ready to receive influences from without.

There is, however, one intimation at an early period of such a population being found in Thrace, from which we may gather that they had migrated further southwards to avoid the tide of invasion. It is found in a curious story, told by Theophanes at the end of the sixth century. At that time a Chan of the Avars had overrun the Eastern empire, and after a series of victories, appeared before the walls of Constantinople. In the meanwhile two of the Roman Generals, who had concealed themselves in the fastnesses of the Balkan, succeeded in mustering a considerable body of troops, and were on their way to surprise the rear of the Avars, when their project was brought to an end by the following occurrence. One of the beasts of burden happened to fall down in the line of march; on which some one close by called out to its driver, in the language of the country, *Torna, torna, fratre,* that is, "Turn him round, brother." The driver did not hear this, but the other soldiers did; and thinking the enemy were upon them, and that this was the sign for retreat, they took up the cry *torna, torna,* and the whole force fled precipitately. The words might be either Italian or Wallachian, but as Italian soldiers were no longer employed in the armies of the Eastern empire, it seems highly probable that the men who used them were Wallach inhabitants of the Balkan.[10]

Although the Byzantine writers were aware that a distinction of race existed between the Greeks and the inhabitants of the mountains of Thrace and Macedonia, the name of Wallachs ($\beta\lambda\acute{\alpha}\chi οι$) does not occur until the year 1027, when they are mentioned by Lupus Protospatha as serving in the Byzantine army, which was then sent out to conquer Sicily. Later on, in the same century, in the reign of Alexius Comnenus, those who dwelt in the hilly country near Constantinople were well known as a source of recruits for the Imperial forces, being hardy mountaineers, inured by long exposure in their occupation of

---

[10] Thunmann, 'Untersuchungen,' p. 341.

shepherds and hunters. Two centuries later (1282) we hear of the same branch of the tribe as having become so numerous and wealthy, as to be a source of fear to the inhabitants of the city. In consequence of this the emperor Andronicus II. took precautions to get rid of them, and transplanted the whole people to Asia Minor, where their numbers were greatly reduced by ill-treatment and the severity of the climate; until at last the remnant were permitted to purchase, with a heavy sum of money, their return to their native soil.

Meanwhile the Wallachs of the Balkan had experienced a separate fortune, and with them the race rose to distinction on the only occasion when they come prominently forward in history. After being subdued by the Bulgarians, and again brought under the Eastern empire, when that nation was subdued by the emperor Basil II., they maintained themselves in their mountain fastnesses, owing an allegiance more or less qualified to Constantinople. In the reign of Isaac Angelus (1186), however, when they were heavily taxed, robbed of their cattle, and misused in other ways, they rose, under the leadership of three brothers, Peter, Asan, and John, and having made a league with the Bulgarians, raised the standard of revolt, and established what is called the Bulgaro-Wallachian kingdom. Its successive rulers contended with varied fortune against the Byzantine government, but succeeded in maintaining their position in Thrace and Macedonia, to which countries for a time Thessaly also was added, forming, however, an independent province, with a governor of its own. The emperor Baldwin, the first of the Latin emperors of Constantinople, was captured by them in battle, and put to death. The kingdom continued to exist until the Turks made their appearance on the scene, when in common with the other independent governments in these regions, it was finally overthrown. Its first founders, out of opposition to Byzantine influence, embraced the religion of Rome, and introduced the Latin worship. When, however, the empire passed into the hands of the Latins, a counter-opposition prevailed, and in order to establish a connection with the rival Eastern emperor at Nicæa, they adopted the Greek rite, to which they have ever since adhered.

That part of the race which occupied Thessaly is sufficiently interesting to deserve an independent notice. Instead of being restricted, as they are now, to a few localities in the chains of Olympus and Pindus, for several centuries they held all the mountains that surround the Thessalian plain, and for a time, as it would appear, even the plain itself. In consequence of this, the usual name for this district in mediæval writers is Great Wallachia (Μεγάλη Βλαχία), in contradistinction to Ætolia and Acarnania, which were called Lesser Wallachia. Just before the establishment of the Bulgaro-Wallachian kingdom, in the year 1170, the Jewish traveller, Benjamin of Tudela, passed through the country; and so great was the impression this people produced on him, that whereas he usually confines himself to short notices of all but the Jewish communities along his route, here he becomes communicative beyond his wont. On reaching Zeitun, the ancient Lamia, near the Maliac gulf, he says:—"Here we reach the confines of Wallachia, the inhabitants of which country are called Vlachi. They are as nimble as deer, and descend from their mountains into the plains of Greece, committing robberies and making booty. Nobody ventures to make war upon them, nor can any king bring them to submission, and they do not profess the

Christian faith. Their names are of Jewish origin, and some even say that they have been Jews, which nation they call brethren. Whenever they meet an Israelite, they rob, but never kill him, as they do the Greeks. They profess no religious creed."[11] In the succeeding period the district which they occupied passed in part, at all events, into a variety of hands; at one time being held by a Western count in dependence on the Latin emperor of Constantinople, at another by one of the despots of Epirus, at another by one of the Greek emperors of Nicæa. But all along until the Turkish conquest a native Wallach governor seems to have existed among them, and to have been in reality supreme.

From that time to the present the Wallachs in Turkey can hardly be said to have had a national existence. They have been subservient members of the Greek Church, and have proved a willing instrument in the hands of the Greeks to assist in checking any expressions of independence on the part of the Bulgarians or other Christian races. In some places, as at Metzovo, the men have even learnt to speak Greek, though in their families they retain the use of their native tongue. But though their numbers, even at the present day, are not insignificant (they are supposed to amount to 400,000), yet there is no need to take them into account in providing for the political future of the Turkish empire.

The present state of the people is well described by Fallmerayer in the following passage, which is corroborated by the accounts of other writers:—

"The Wallachs of Thessaly, equally with their relations in speech and race in the Danubian Principalities, call themselves Romans.

---

[11] Benjamin of Tudela, in Bohn's 'Early Travels in Palestine,' p. 72.

They speak a corrupt Italian, and have their head-quarters on the ridge and the declivities on both sides of the Pindus, in the valleys which give birth to the Peneius and the neighbouring rivers, where they are first mentioned by the Byzantine historians of the eleventh century. They hold and command the gates between Thessaly and Albania; and Metzovo, a stone-built town of about 1000 houses, on the chain which forms the division between the defiles that descend in opposite directions, may be regarded as their chief seat. Malacassi and Lesinitza, but especially Kalarites,[12] Kalaki, and Klinovo, with more than twenty other villages in and about the gorges of Pindus, belong also to this people, who, on account of the rude climate of their native place, are but slightly occupied in husbandry, but devote themselves all the more successfully to the breeding of cattle and sheep-farming on the most extensive scale, and have won for themselves a position of importance throughout Rumelia by their wealth in flocks. In the winter season, when the mountain-heights are covered with snow, they lead a nomad life and pasture on the grassy plains of the temperate lowlands, even as far as Greece, until the return of spring drives the black encampments of the wandering Wallach shepherds back to the mountain-pastures.

"The Wallachs are as superior to the Greek-speaking population in sobriety, economy, and industry as they are inferior to the Greco-Slavs in refinement of manner, in spirit, and in general shrewdness. Nevertheless, these simple and rude shepherds are remarkably skilful in metal-work. The weapons and accoutrements inlaid with gold and silver, which we admire on the persons of the Arnaouts and Palikars, are produced in the workshops of the Wallachs; while the waterproof capotes, which under the names of *Capa, Greco*, and *Marinero*, are well-known in the ports of the Mediterranean, are for the most part to be regarded as the product of Wallach woollen manufacture. Wallach shopkeepers and artizans are to be found in all the cities of European Turkey, and they are led by the love of money-making even as far as Hungary and Austria. That they also understand business on a large scale is shewn by the wealthy Sina at Vienna, who is by birth a Wallach of Klinovo, if we mistake not, or at all events comes from one of the above-named places in Pindus. But, like all other mountaineers,

---

[12] Of this place, which is situated among the mountains, at some distance to the south-west of Metzovo, an interesting account is given in Leake's 'Northern Greece,' i. p. 274.

the Wallach never forgets his home in the most distant country; and in very numerous instances he returns in old age to the Pindus with the fruits of the labour of his life, in order that he may rest in the same soil with his fathers." [13]

To this account it should be added, that they are the great carriers of this part of Turkey, having in fact almost a monopoly of this branch of trade between Thessaly and Epirus.

[13] Fallmerayer. 'Fragmente aus dem Orient,' ii. pp. 240, seq.

CHAPTER XXVI.

YANINA AND ZITZA.

Valley of the Arta — Races of Southern Albania — Yanina and its Lake - Oracle and Sanctuary of Dodona — Excursion to the Monastery of Zitza — Byron's Description of it — Albanian Rifle Practise — Scene of Ali Pasha's Death — His Serai and Tomb — Notice of his Character and Career — Atrocious Massacre of Greek Women.

In the afternoon of the same day we left the khan opposite Metzovo, and rode along the side of a deep gorge, at a considerable height above the rushing stream of the Arta. When at length we descended to the river, we found that our path lay through its bed for several hours, during which we crossed and recrossed it not less than twenty-three times. The wooded mountains which rose on either side were fine, and there was a grand breadth about the scenery generally, in which respect it recals the valleys of the western Tyrol, and like them, after a time becomes monotonous. At length the river makes a bend to the southward, in which direction it flows through the remainder of its course to the Gulf of Arta (Ambracian Gulf); between it and the basin of Yanina lies the steep ridge of Drysco, or "Oaklands," over which the road has to pass. At the point where it leaves the river there is a solitary khan, shaded by a fine tree, which we recognised as the same for sitting under the shadow of which Mr. Curzon amusingly describes that he was charged in his bill, in default of any other means of extortion. Throughout these parts we found the Romaic language still spoken, and when you emerge from among the Wallachian population you find yourself once more in the midst of Greeks. On this point it may be well to say a word, now that we are entering Southern Albania or Epirus, because considerable confusion has existed in some travellers' minds about it, and this has led to serious mistakes. In our journey through Central Albania, when we arrived at the valley of Argyro-Castro, we observed that it was the meeting point of two nationalities, the northern half being occupied by Albanian Tosks, while the southern part, and the districts beyond it, are inhabited by Greeks. Now the province of southern Albania, that is the country south of the parallel of Argyro-Castro, may be regarded as divided in two parts in respect of race, by a line drawn at first south-eastwards in the direction of Yanina, and then directly southwards to Prevyza at the mouth of the Gulf of Arta. To the east of this line, including all the neighbourhood of Yanina, and the valleys of the Arta and Aspropotamo, the population is purely Greek; to the west it is composed of two elements, Greek and Albanian, and the people speak both those languages, though one or the other is more familiar to them, according to the predominant element in their race. Thus the Suliotes, whose territory was the knot of mountains in the southern part of this division, and who are so famous for their heroic resistance to Ali Pasha, were pure Albanians, and in their families spoke only the Albanian language. In this case the confusion has been the greater, because, from their having been regarded as Greeks, their deeds of valour have often been introduced in connection with Hellenic intrepidity and patriotism.

When we reached the summit of the ridge of Drysco a magnificent view burst on us of Yanina and its lake, enclosed by a wide circle of fine mountains. The lake is about six miles in length and two in breadth; on the eastern side of its bright expanse rises the grand mass of Mount Metzikeli, which reaches the height of 3000 feet above the sea, while on the western shore lies the city itself, with its citadel built on a conspicuous rock, which projects some distance into the water. Behind it the buildings rise gradually up a sloping hill-side, shining domes and minarets, dark roofs and foliage being picturesquely combined, in the way that is so characteristic of Turkish towns. To the south of the lake a wide marshy plain extends, broken by the long craggy ridge of Castritza, and the distant grey peaks of Suli bound the view. We descended along a road constructed by Ali Pasha, parts of which we had already passed at several points in the valley of the Arta: it had been originally built of massive materials, but being now rough and broken, it only served to show the one line to be avoided. When we reached the southern end of the lake we had to make a considerable détour to avoid the reedy swamps at its foot, from which the hill of Castritza rises. Close under that eminence Colonel Leake discovered the subterranean passage, or catavothra, by which the water escapes, and which here bears the expressive name of "the digester" ($\chi\omega\nu\epsilon\acute{\nu}\tau\rho\alpha$). On reaching the opposite side we made our way along a turfy level to the city, which we reached about dusk, having ridden nearly sixty miles in the day.

Joannina, or St. John's Town, which is now almost always known as Yanina, or Jack's Town, contains at present between 22,000 and 25,000 inhabitants, having greatly fallen off in this respect since the time of Ali

Pasha, when its population amounted to 35,000. The Greek Christians are nearly three times as numerous as the Mussulmans; and there is also a considerable number of Jews. From its height above the sea, which is estimated at 1000 feet, and the neighbourhood of lofty mountains, the temperature in summer is less oppressive than that of most cities in Turkey; and though the lake, like many of those in Greece, varies in extent with the seasons, and consequently is fringed with marshes instead of a pebbly beach, yet for ordinary residents the place is said not to be unhealthy. The lower parts of the town, however, are an exception to this, for in them, as might be expected, dysentery and malaria fever prevail to a great extent during the summer. To sight-seers the principal object of attraction are the bazaars, in which may be seen specimens of the rich gold embroidery for which the place is famous; here also the dresses of the inhabitants are displayed to the greatest advantage, being the same elaborate costume which the modern Greeks have adopted since they obtained their independence. The most conspicuous object from every part of the town is Mount Metzikeli, whose gigantic precipices of grey limestone, seamed by the courses of numerous torrents, appear to rise immediately from the water on the opposite side, and when darkened, as they so frequently are, by a cap of thunder-clouds, seem extraordinarily near.

In modern times this place has mainly been celebrated as the residence of Ali Pasha's magnificent semi-barbaric court, and the chief scene of his intrigues, his atrocious crimes, his resistance to the power of the Sultan, and at last of his tragical end. But to the classical student it presents even greater interest as being the probable site of Dodona, the great Oracle of

Zeus, and the head-quarters of the Pelasgian race in this their early home, second only to Delphi among all the sacred places of Greece. No traces, it is true, have hitherto been found of the temple or its substructions, and therefore we cannot speak with any certainty on the subject; but Colonel Leake, who has discussed the question with his wonted learning and acuteness, has proved thus much—that it was undoubtedly situated in this neighbourhood, and that no place can be found so suitable for it as Yanina. He himself inclines to the opinion that the *city* of Dodona was situated on the height of Castritza at the southern end of the lake, where extensive remains of ancient walls are found, but that the *oracle* and sanctuary occupied the rocky promontory on which the castle of Yanina now stands. A brief recapitulation of the arguments by which he arrives at this conclusion has been given at the end of this volume:[1] assuming, however, that his view is the true one, let us listen to his description of the appearance of the ancient sanctuary, as far as it can be gathered from a comparison of notices in ancient writers.

"In place of the dirty streets and bazaars of the modern town, we may imagine a forest, through which an avenue of primæval oak and ilex conducted to the sacred peninsula. Within the porticos which enclosed the temple were ranges of tripods supporting cauldrons, the greater part of which had been contributed by the Bœotians in consequence of an annual custom, and which were so numerous and so closely placed that when one of them was struck the sound vibrated through them all. Many others had been dedicated by the Athenians, whose Theoria, or Sacred Embassy, brought yearly offerings; but the most remarkable of the anathemata was a statue dedicated by the Corcyrei, holding in its hand a whip with three thongs loaded with balls, which made a continual sound as they were agitated by the wind against a cauldron. In a picture of the temple of Dodona, which has

---

[1] See Appendix G, "On the Site of Dodona."

been described by Philostratus, the prophetic oak was seen near the temple, and lying under it the axe of Hellus, with which he struck the tree, when a voice from it ordered him to desist. A golden dove, representing the bird of Egypt which uttered the voice, was perched upon the tree; garlands were suspended from its branches, and a chorus from Egyptian Thebes was dancing around it, as if rejoicing at the recognition of the sacred dove from their native city. The Selli [the priests of the temple] were seen employed in prayer or sacrifice, or in decorating the temple with fresh boughs and garlands, or in preparing cakes and victims, while the priestesses were remarked for their severe and venerable appearance. Whether this be the description of a real picture, or the ideas of Philostratus for the subject of one, it is probably a faithful portrait of the hierum of Dodonæan Jove, in the height of its reputation, when it may easily be supposed that the temple, the porticos, the dedications, and the dwellings of the sacred servants were sufficient to occupy the greater part of the peninsula."[2]

The morning after our arrival at Yanina we paid a visit to the English vice-consul, Signor Damaschino, who gave us some useful information about the country we were now going to explore, and promised us introductions to the commander of one of the castles of Suli, and to persons in some of the remote villages which we should pass on the way. When we had taken leave of him we started about midday on an expedition to the monastery of Zitza, about fifteen miles off to the northwest, which elicited such noble verses from Byron's pen, in his description of its position and the view it commands, in the second canto of 'Childe Harold.' We had now sent back the post horses which came with us from Larissa, as the *menzil* comes to an end here, and had to hire others from a *kharidji* or carrier for the rest of our journey. During the first part of the way the track lay through the meadows at the side of the lake of Yanina, after leaving which we passed another smaller piece of water—the lake of Lapsista—which is also

[2] Leake, 'Northern Greece,' vol. iv. pp. 198, foll.

supplied from sources at the foot of Mount Metzikeli. The subterranean channel, by which its waters disappear, is entered by a number of small mouths on its western side close to the road. The valley of Yanina, which is partly filled up by these waters, is altogether twenty miles in length from north to south, and has no outlet for its waters at either end. On the western side the rich pastures extend for a considerable distance up the mountain slopes. At length we struck up along the hills to our left, until we saw the village of Zitza before us, just above which, on an eminence, stands the small white-walled monastery, charmingly situated amidst a thick grove of oaks and elms. It is a dilapidated old place, though now partly restored; we entered by a battered iron door, and met with a hearty welcome from the assembled fathers—five in number—one of whom, a venerable old man, said he remembered Byron's visit.

They conducted us to a pleasant spot beneath the trees, from which we could enjoy the view, and maintained the character which Byron has given them of being "not niggard of their cheer," by helping out our meal with some very palatable white wine, which comes from the vineyards that surround the monastery. They told us that it is partly pressed by the feet, partly by the hand; the latter method, which is not usual, is mentioned in one of the notes to 'Childe Harold.' As we reclined there a dull murmur reached our ears, proceeding from a waterfall formed by the stream of the Calamas, the ancient Thyamis, which flows a little way off to the west. As to the scenery—though Byron's judgment in this respect is so good as hardly to admit of question, yet in this instance I cannot help thinking that he overrated what he saw. As he was all but lost in a storm in the neighbourhood of this monastery he was in all probability more than usually disposed to appreciate the place where he found refuge, and all its surroundings. Besides this it is likely enough that the higher mountains were covered with snow at that time, an element which greatly enhances the beauty of every scene. It is a very extensive view, and comprises magnificent mountain chains, but there is a want of colour, and very little variety, nor are the different objects pleasingly arranged: one long line of table-land in particular, half mountain half plain, which stretches away in the direction of Yanina, and excludes that city from view, is anything but agreeable to the eye. Here, as in most of the scenery west of the Pindus, there is but little of that classical beauty of sharply-cut outline, and that finely-balanced grouping of the component parts in each view, which are so characteristic of the mountains in the rest of the Greek peninsula. Let me now introduce Byron's highly-wrought description:

> "Monastic Zitza! from thy shady brow,
>   Thou small, but favour'd spot of holy ground!
>   Where'er we gaze, around, above, below,
>   What rainbow tints, what magic charms are found!
>   Rock, river, forest, mountain, all abound,
>   And bluest skies that harmonise the whole:
>   Beneath, the distant torrent's rushing sound
>   Tells where the volumed cataract doth roll
> Between those hanging rocks, that shock yet please the soul.
>
> "Amidst the grove that crowns yon tufted hill,
>   Which, were it not for many a mountain nigh
>   Rising in lofty ranks, and loftier still,
>   Might well itself be deem'd of dignity,
>   The convent's white walls glisten fair on high:
>   Here dwells the caloyer, nor rude is he,
>   Nor niggard of his cheer; the passer-by
>   Is welcome still; nor heedless will he flee
> From hence, if he delight kind Nature's sheen to see.

> "Here in the sultriest season let him rest,
> Fresh is the green beneath those aged trees;
> Here winds of gentlest wing will fan his breast,
> From heaven itself he may inhale the breeze:
> The plain is far beneath—oh! let him seize
> Pure pleasure while he can; the scorching ray
> Here pierceth not, impregnate with disease:
> Then let his length the loitering pilgrim lay,
> And gaze, untired, the morn, the noon, the eve away."

To this the poet adds the following note:—

"The situation is, perhaps, the finest in Greece; though the approach to Delvinachi and parts of Acarnania and Ætolia may contest the palm. Delphi, Parnassus, and, in Attica, even Cape Colonna and Port Raphti, are very inferior, as also every scene in Ionia or the Troad: I am almost inclined to add the approach to Constantinople, but, from the different features of the last, a comparison can hardly be made."

As we were returning, just after sunset, when it was becoming dusk, one of my companions had ridden on in front of our party, and after rounding the projecting knoll of a slope, along which lay our track, drew in for a minute to await our arrival. Just as we came up a flash and bang proceeded from some thick brushwood not far off, and he heard the *ping* of a bullet, which passed close to him. No harm was done, and we heard nothing more of the marksman; but it served as a salutary hint as to the sort of sport that occasionally goes on in Albania, and the benefit of being tolerably sharply on the look out. It was the only occasion on which I have been conscious of any risk in travelling in these countries, and I do not believe, if reasonable precautions are taken, that there is any real danger.

The next morning we visited the scene of Ali Pasha's death. This is on a small rocky island in the lake opposite the castle, but nearer to the other shore under Mount

Metzikeli. The boats which are used on this lake are called *monoxyla*, but they are not scooped out of one piece of wood, as the name would seem to imply, and as is the case with some of the river-boats in Turkey, but are built of thin planks, running very fine fore and aft, and are propelled at a considerable pace with broad paddles. Stretched at the bottom of one of these, we glided across to the island, and, rowing round to the opposite side, disembarked near some mean cottages, where, in a sheltered nook beneath a rock, and almost entirely shaded by an immense plane-tree, appeared the little monastery of St. Panteleemon, the object of our visit. It was to this spot that Ali retired at the end of his career, when he was now seventy-four years of age, having first been driven by the forces of the Sultan within his capital, and then forced to evacuate it, after witnessing the destruction of the fortifications and many of the public buildings of the place in the course of the siege. We entered by a low-arched door, and ascended to a small balcony, upon which open two or three chambers. In this balcony he was invited to a conference with some emissaries of the Sultan, by whom he had been led to expect that terms would be granted him. One of these, Kourchid Pasha, waiting for a moment when the old man's watchful eye was averted, treacherously stabbed him; but, not being mortally wounded, he managed to retire into one of the small rooms behind. The Sultan's troops then entered the place, and the Albanians, who commanded the court from the cliffs above, being prevented from firing on them by the presence of one of their comrades who was within, the hostile soldiery made their way into the chamber below where Ali was, and, firing through the wood floor, succeeded in killing him. The holes in the floor where the bullets passed through are still to be seen. His favourite wife, Vasilike, was in the adjoining room at the time of his death, and if the monster's orders had been carried into execution she would have been put to death immediately after; happily, however, her life was saved. These events took place on the 5th of February, 1822.[3] His head was then cut off and sent to Constantinople, where it was exposed at the gate of the Serai, together with those of his three sons and one of his grandsons, who had been permitted to live in Asia Minor until their father's overthrow, but were then put to death. They were buried in the great cemetery outside the walls of Stamboul, where five turbaned head-stones, ranged in a line by the road-side, attract the attention of the passer-by.

As we returned, the massive towering walls of the ruined citadel presented a grand appearance, rising from their foundation of solid cliff, with the glassy mirror of the lake beneath. There we landed for a short time to see Ali's serai, and the tomb which covers his headless trunk. The former of these is now a somewhat ruinous-looking building, and the painting with which the walls are decorated has an unsightly appearance. The monument, which is an unpretending tomb of white marble, is fenced in by ornamental iron-work, resembling a large bird-cage, and stands in the open space in front of the palace. Close by this is the mosque where Ali used to worship. The citadel is separated from the mainland by a wet ditch of artificial construction, over which a drawbridge is thrown.

[3] Mr. Finlay, in his 'History of the Greek Revolution' (i. p. 116), has given a somewhat different account of Ali's death; but the story given above is that which is known on the spot, and has been related before by other travellers.

The career of Ali Pasha is too well known to require to be narrated here, but the sketch of his character which Mr. Finlay has given in his 'History of the Greek Revolution,'[4] is so true and so effective that I am tempted to quote it. He introduces it with the pertinent remark that, in Albania, as in Greece in the time of Homer, no genealogy is carried by name beyond the great-grandfather of the most distinguished man. Thus, though his family is known to have been long settled at Tepelen, yet the first of whom we hear was his great-grandfather Mutza Yussuf, or Moses Joseph, who raised himself to considerable power by his personal valour. His son and successor, Mukhtar Bey, was slain at the siege of Corfu, fighting against Schulemburg. Veli, the third son of Mukhtar, was accused of poisoning his two elder brothers to secure the chieftainship, and himself also died young.

"Ali, the infant son of Veli, was left to the care of his mother, whose relationship to Kurd Pasha, of Berat, a powerful Albanian chieftain, secured protection to the infant. The young Ali grew up in lawless habits. Sheep-stealing involved him in local feuds, and, falling into the hands of an injured neighbour, he was only saved from death by the interference of Kurd Pasha. He then entered the Sultan's service, and was employed by Kurd as a guard of the dervens. He was brave and active, restless in mind and body, and utterly destitute of all moral and religious feeling: but his good-humour made him popular amongst his companions. He displayed affection to the members of his family and gratitude to his friends. As he grew older and rose in power, he became, like most Albanians, habitually false; and regarding cunning as a proof of capacity, his conversation with strangers was usually intended to mislead the listeners. During his long and brilliant career his personal interests or passions were the sole guides of his conduct. Within the circle of Albanian life his experience was complete, for he rose gradually from the position of a petty chieftain to the rank of a powerful prince; yet his moral and political vision seems never to have been enlarged, for at his greatest elevation

---

[4] Vol. i. pp. 71, *seqq.*

selfishness obscured his intellect and avarice neutralised his political sagacity. His ambition was in some cases the result of his physical activity.

"Ali, like every Albanian or Greek who has risen to great power by his own exertions, ascribed his success solely to his own ability, and his self-conceit persuaded him that his own talents were an infallible resource in every emergency. He thought that he could deceive all men, and that nobody could deceive him; and, as usually happens with men of this frame of mind, he overlooked those impediments which did not lie directly in his path. As an Albanian, a pasha, and a Mohammedan, he was often swayed by different interests; hence his conduct was full of contradictions. At times he acted with excessive audacity, at times with extreme timidity. By turns he was mild and cruel, tolerant and tyrannical; but his avarice never slept, and to gratify it there was no crime which he was not constantly ready to perpetrate.[5]

"The boasted ability of Ali was displayed in subduing the Albanians, cheating the Ottoman Government, and ruling the Greeks. His skill as the head of the police in his dominions gave strangers a favourable opinion of his talents as a sovereign. He found knowledge useful in his servants: he therefore favoured education. His household at Joannina had all the pomp and circumstance of an Eastern court; but it had no feature more remarkable than a number of young pages engaged in study. The children of Albanian Mussulmans might be seen in one

---

[5] Colonel Leake, writing from Yanina at the time when Ali was at the height of his power, says:—"The Pasha's avaricious disposition carries him to such a length that he never allows any worn-out piece of furniture, or arms, or utensils, to be thrown away, but lays them in places well known to him, and would discover the loss of the smallest article. In the dirty passages and ante-chambers leading to some of the grandest apartments of his palace, and which have cost some thousands to fit up, the worn-out stock of a pistol, or a rusty sword, or a cabbard, or some ragged articles of dress, may be seen hanging up, which his numerous domestics never venture to remove, well knowing that it would be remarked by him. This mixture of magnificence and meanness is very striking in every part of the palace. His great apartment, covered with a Gobelin carpet, surrounded with the most costly sofas, musical clocks and mirrors, is defended by cross iron bars, rougher than would be considered tolerable in the streets of London. They are intended to prevent his servants from passing through the window when the chamber is locked."—'Northern Greece,' i. pp. 405, 6.

ante-chamber reading the Koran with a learned Osmanlee, while in another room an equal number of young Christians might be seen studying Hellenic grammar with a Greek priest.

"Under Ali's government Joannina became the literary capital of the Greek nation, for he protected laymen who rebelled against the Patriarch and Synod of Constantinople, as well as priests who intrigued against the Sultan. Colleges, libraries, and schools flourished and enjoyed independent endowments. He ostentatiously recommended all teachers to pay great attention to the morals of their pupils, and in his conversation with Greek bishops he dwelt with a cynic simplicity on the importance of religious principles, showing that he valued them as a sort of insurance against dishonesty and a means of diminishing financial peculation. Greek, being the literary language of Southern Albania, was studied by Mussulmans as well as Christians. Poems and songs, as well as letters and accounts, were written by Mohammedans in Greek, and many were circulated in manuscript. Unfortunately, no collection of Mohammedan songs and poems has been published.[6]

"The cruelty of Ali excited horror in civilised Europe, but it extorted admiration from his barbarous subjects. The greatest compliment they could pay him was to praise his cruelty to his face. Persons still living have seen him listen with complacency to flattery embodied in an enumeration of his acts of direst cruelty, and shuddered at his low demoniacal laugh when his Greek secretaries reminded him how he had hung one man, impaled another, and tortured a third. Lord Byron might well say, that—

'With a bloody hand
He ruled a nation turbulent and bold.'"

His most celebrated deed of cruelty deserves to be mentioned, on account of the frightful atrocity of the circumstances by which it was accompanied. It originated in the jealousy of his daughter-in-law, whose

---

[6] Lord Byron remarks, in one of the notes to 'Childe Harold':— "'Athens,' says a celebrated topographer, 'is still the most polished city of Greece.' Perhaps it may be of Græce, but not of the Greeks, for Joannina, in Epirus, is universally allowed, amongst themselves, to be superior in the wealth, refinement, learning, and dialect of its inhabitants."

The colleges referred to above, which were endowed with private benefactions, have now dwindled away almost to nothing.

husband Mukhtar, Ali's eldest son, had been attracted by the charms of Euphrosyne, a Greek lady of great beauty and attractive manners, but graceless conduct. At her instigation he determined to redress her wrongs, professing at the same time the intention of eradicating a great social evil, and reforming the morals of his city. It was said, however, that it was in reality an act of private revenge on his part, in consequence of his having been an unsuccessful suitor to this lady; and the notorious profligacy of his life rendered the charge sufficiently probable. The scheme of a general massacre of all the women who were suspected of criminality was conceived with the greatest deliberation, and carried into effect in the following manner:—

"Ali was in the habit of dining with his subjects at their own houses when he wished to confer on them an extraordinary mark of favour. He signified to Nicholas Yanko, whose wife was one of the proscribed, his intention to honour him with a visit. The men dine alone in Eastern lands. After dinner, the great Pasha requested that the lady of the house might present his coffee, in order to receive his thanks for the entertainment. When she approached, he addressed her in his usual style of conversation with Greek females, mixing kindness with playful sarcasm. Rising after his coffee, he ordered the attendants in waiting to invite several ladies, whose conduct, if not virtuous, had certainly not been scandalous, to visit Yanko's wife at her house.

"Ali proceeded to the house of Euphrosyne, attended by a few guards, and, walking suddenly into her presence, made a motion with his hand, which served as a signal for carrying off the victim, who was conveyed to Yanko's house much more astonished than alarmed. Ali rode on to his palace, and engaged in his usual employments. The ladies of the party assembled at Yanko's house were soon discomposed by having an equal number of females of the very lowest order in Joannina thrust into the room by policemen. In a few minutes the whole party was hurried off to the church of St. Nicholas, Yanko's patron saint, at the northern extremity of the lake. There the unfortunate culprits were informed that they were condemned to death by the Pasha. The wealthier were at first not much frightened; for Ali's avarice was so notorious that they believed their relations would either voluntarily ransom their lives, or be compelled to do so. The worst punishment they feared was imprisonment in the convents on the islands of the lake.

"Morning had dawned before the party reached the church of St. Nicholas, and Mohammedan customs require that the execution of a sentence of death on females by drowning must be carried into effect while the sun is below the horizon. For twenty hours ladies of rank and women of the lowest class remained huddled together, trembling at times with the fear of death, and at others confident with delusive hopes of life. At sunset a violent storm swept the surface of the lake, and it was midnight before they were embarked in small boats and carried to the middle of the lake. There they were thrown overboard, without being tied up in sacks, according to the Mussulman formality in executing a similar sentence. Most of the victims submitted to their fate with calm resignation, sinking without an audible word, or with a short prayer; but some resisted to the utmost, with piercing shrieks; and one, whose hands had got loose, clung to the side of the boat, and could only be plunged under water by horrid violence. When all was finished, the police guards watched silently in the boats until morning dawned; they then hastened to inform the Pasha that his orders had been faithfully executed. One of the policemen present, who had witnessed many a horrid deed of torture, declared, long after, that the scene almost deprived him of his senses at the time, and that for years the voices of the dying women were constantly echoing in his ears, and their faces rising before his eyes at midnight."[7]

---

[7] Finlay, 'History of the Greek Revolution,' i. pp. 75-77. The account given by Colonel Leake, who derived it from Yanko himself, adds some further details of the tragedy.—'Northern Greece,' i. pp. 401-4.

CHAPTER XXVII.

SULI.

Frequency of Thunderstorms in Epirus — Greek Theatre at Dramisius — Ruins of an Hellenic City — Mount Olytzika — The Molossian Dog — Sources of the Charadrus — Mountains of Suli — Moonlight Views — Kako Suli — Ruined Homes of the Suliotes — Their History — Invasion of their Country by Ali — Samuel the Caloyer — Their Expulsion — Gorge of the Acheron — The Palus Acherusia — The Inferno of the Greeks — Parga — Its Cession to the Turks — Ballad on the Subject.

WE had now arrived at the 18th of June, and on the afternoon of that day we made a fresh start, being bound for the mountain district of Suli, and the scene of the struggle of the Suliotes. In the course of our journey we had plentiful proof of the frequency of atmospheric changes in this neighbourhood of the "thunder-hills of fear," which caused it to be a suitable home for the worship of Jupiter Tonans. During the morning, thunderstorms had been hanging about Mount Metzikeli, and lent an additional grandeur to the gloom of its precipices, but now they had gathered on all the heights around, and we could see the forked flashes glittering on many sides. As we passed the craggy ridge of Castritza, which was now on our left hand, the lightning that played among its peaks, the peals of thunder, and the thick clouds floating round the distant mountains, added a wild charm to the whole scene. In the midst of this we took shelter for a short interval at a khan, but, as time pressed, we braved the storm again, which descended upon us in torrents of rain, until towards evening the murky atmosphere cleared away, and the sky was filled with mellow light. We had passed over the plain at the lower end of the lake, and, after crossing a range of hills, descended into a broad valley, the southern end of which is bounded by the vast mass of Mount Olytzika, whose lofty grey cliffs form an outwork to the stronghold of Suli. This mountain, which is the most central and the most imposing of all in this neighbourhood, may be regarded as a point of demarcation of the races which have been already mentioned as inhabiting Southern Albania, the country to the east of it being occupied by pure Greeks, while that to the west is the abode of mixed Greeks and Albanians, though in this part the latter are the predominant race.

Just before reaching the level land near the village of Dramisius, we came upon a magnificent ruin,—the finest Greek theatre now existing, with the sole exception of that of Dionysus at Athens, which has been excavated within the last four or five years. It is placed on the side of a hill, just where it descends to the plain; so that, like most other Greek theatres, it is supported by the ground at its back, and has no need of arches or other masonry, except at the ends of the cavea, where extensive walls and fine buttresses still remain. The form is a semicircle, somewhat elongated; but this peculiarity is not so distinctly marked here as in most of the theatres of continental Greece; the main point of difference between the Greek and Roman theatre in respect of form being that, while the latter is an exact semicircle, the former is elongated in the direction of the *scena*. A distinction is also to be drawn in this respect between those of Greece and those of Asia Minor, namely, that in the Asiatic colonies the cavea assumes a horse-shoe shape, while in the mother-country the elongated sides are always parallel to one another. Its size is very large; in fact, it is the largest in Greece, with the exception of that at Sparta, and probably also of that of Dionysus at Athens. The exterior diameter is given by Leake as 445 feet, while that of the theatre at Sparta is 453 feet; several, however, in Asia Minor exceed this size.[1] The seats, which are composed of a fine white limestone, nearly approaching to marble, almost all remain, but, owing to the dislocations produced by earthquakes, and by the shrubs which for ages have grown amongst them, they are thrown out of their places in the most extraordinary way, and so make the place appear more ruinous than it really is. In consequence of this, it is less easy than in some other theatres to trace the *diazomata*, or landing-places, which ran at intervals round the building, thus dividing it into separate tiers; and the flights of steps by which the spectators reached their seats. On the lowest level towards the plain, beyond the *cavea*, other foundations are visible, in a line with which the *scena* itself must have been; but of this and the *proscenium* there are no remains. The theatre commands a fine view of the hill of Olytzika opposite, and of the deep valley which runs up beneath its eastern flank; so that it adds another to the numerous instances of the good taste of the Greeks in their choice of positions for temples and theatres.

The existence of this magnificent building in so remote a position shows how completely the wild region of Epirus must have been pervaded by Hellenic influences. In many respects these scenic edifices are amongst the most interesting and instructive remains that have come down to us from antiquity. They enable our mind's eye to reproduce the people, when congregated together in

---

[1] Leake's 'Tour in Asia Minor,' pp. 328-9.

the greatest numbers, animated by the strongest sense of enjoyment, and excited by the most varied emotions. They impress us also with an idea of the wealth of the communities which could erect structures of such beauty and such durable materials; and of the religious feeling, combined with love of art, which urged them to celebrate performances on such a scale in honour of the gods. And, whatever may have been maintained by Hume and other writers with regard to the limited population of ancient states, and the causes which tended to check their increase, the vast size of these buildings, which enabled them in many cases to accommodate from ten to twenty thousand spectators, is a proof, far more convincing than figures, of the teeming multitudes which must have crowded the cities.

Behind the theatre, and at one point nearly touching it, there are fine remains of the walls of an Hellenic city, or, perhaps, as the area which they comprise is very confined, of a sacred enclosure. The blocks of which these are composed are partly laid in horizontal courses, and partly fitted into one another with great nicety at irregular angles. Within one part of the *enceinte* there is a curious subterranean building of Greek construction, to which we were conducted by some peasants. It was square in form, and of narrow dimensions; and within it two pilasters supported a roof of flat stones, which lay about the level of the earth. Had it not been for its small size, and the position of the rocks at its sides, it might have been taken for the substructions of a temple; as it could not have served that purpose, possibly it may have been a treasury. We have as yet no clue by which to discover what was the ancient name of these ruins. As the place is too insignificant to have been an important town, and disproportionately small to the neighbouring theatre, Leake's supposition is not unreasonable,—that both belonged to a place of public meeting for the Molossian cities, for sacred festivals, and perhaps for civil purposes also.[2] The situation, too, he observes, instead of being strong, commanding, and well watered, the usual requisites of the fortified towns of Greece, is a retired valley like those of Epidaurus, Nemea, and Olympia, which contained similar sacred places. As Plutarch in his life of Pyrrhus mentions a place called Passaron, in the district of Molossia, where the kings of Epirus and their assembled people were accustomed to take mutual oaths, we might at first sight suppose that it corresponded to this position. Unfortunately, however, that place is elsewhere described as having been strongly fortified, and apparently near the sea-coast. Unless, therefore, we accept Dr. Wordsworth's suggestion,—that it is Dodona itself,—which seems to be excluded by some of the arguments which have been adduced relatively to that site,[3]—we must be content for the present to remain in ignorance of its real name and object.

Leaving the ruins late in the evening we proceeded across the valley to our destination, the village of Alipuchori, or Foxtown, which lies directly at the foot of Mount Olytzika. The moon had now risen, and the evening star, which hung in front of us, "shook" from very fulness of light "in the steadfast blue" of such a heaven as can only be seen on a southern evening after rain. On entering the village we were attacked by a number of fierce dogs, which, as we were now in the heart of the ancient Molossia, it was a temptation to regard as the descendants of those Molossian dogs which were so famous in classical times. On another occasion, when

---

[2] 'Northern Greece,' iv. p. 80.　　　[3] *See* Appendix G.

travelling in the valley of Sparta, I had also noticed the extreme ferocity of the dogs in those parts, and called to mind the fame of the dogs of Taygetus. But any conclusion drawn from such observations as these would probably be erroneous. If Lucretius is right in ascribing to the Molossian dogs "magna mollia ricta,"[4] or, as Mr. Munro's photographic translation has it, "large spongy opened lips," they must have been something very different from the lean-jawed, narrow-muzzled animals which are now found with but little variety everywhere throughout the Greek peninsula. Besides this, what I have already observed with regard to the change in the vegetation of these countries, applies to some extent to the animals also. And though there is a difference in this respect between domestic animals, like the dog, and wild ones, whose presence must almost entirely be determined by conditions of climate and food; yet it is highly probable that during the numerous migrations of tribes that passed over the country, a new race of dogs, the necessary accompaniment of nomad hordes, would have been introduced. The wide distribution of the Wallach nation at one period, and their pastoral habits, would in itself have greatly tended to bring about such a change.[5]

[4] Lucret., v. 1063.
[5] On the subject of the ancient breed, my friend Professor Rolleston has favoured me with the following communication:—"I apprehend that some doubt must beset the identification of the *Canis Molossus* of Horace, and Virgil, and Aristotle, with any of the modern varieties, for I see that Fitzinger in his paper, 'Die Raçen des Zahmen Hundes,' which appeared last year in the 'Sitzungsberichte' of the Vienna Academy of Sciences, says that he thinks one of the larger kinds of greyhound was meant by it. Now, for two reasons, I am inclined to differ from this very great authority, whose memoir ought to be consulted by anyone who henceforward writes about the matters of which it treats. First, we know that the *Canis Molossus* was a *barking* dog. This is clear from Horace, whom I need not quote; from Seneca, 'Hippolytus,' v. 30; and from Lucan, 'Bell. Civ.', iv. 440; above all—

'Venator

The intricacy of the paths in this mountain district is so great, and the villagers were so ignorant of the country beyond their own immediate neighbourhood, that we had great difficulty in finding any one to guide us in the direction of Suli, and the kharidji who accompanied our horses we found to be totally unacquainted with the route. Having at length discovered a man who professed

> ' Venator tenet ora levis clamosa Molossi
> Spartanos, Cretasque ligat, nec creditur ulli
> Sylva cani, nisi qui presso vestigia rostro
> Colligit, et præda nescit latrare reperta.'

Secondly, these quotations, and many others which I could give, and others know of, show that the *Canis Molossus* was a *common* dog, a *large* dog, and a dog *good against wild beasts*. Now, certain works of Roman art recently shown me by the well-known antiquarian, J. E. Lee, Esq., and reproduced by his exertions in the Publications of the Antiquarian Society of Caerleon, show that a large dog, good against wild beasts, which the Romans possessed, was a mastiff; and the mastiff barks, which the bulldog and greyhound do not. Hence with the modern naturalists I should call a mastiff *Canis Molossus*, and think it probable that the Romans, and, *mutatis mutandis*, the Greeks, too, did the same. On the other hand, Ennodius (fl. A.D. 500) has the following line as to the figures on the ring of Firmina:—

> Immobilis stantem fugitat lepus arte Molossum.—(Epig. 98.)

But a poet who makes a false quantity cannot be much of an authority either as to words or animals."

Cuvier gives the following accurate description of the modern dogs:—"They are as big as a mastiff, their thick fur is very long and silky, generally of different shades of brown; their tail is long and bushy; the legs seem more calculated for strength than excessive speed, being stouter and shorter than those of the greyhound; their head and jaws are elongated, and the nose is pointed."—'Animal Kingdom,' Griffith's ed., ii. p. 327.

The superiority of the Spartan and Molossian dogs to those found elsewhere in the country, in modern as well as in ancient times, may, perhaps, be sufficiently accounted for by climatic influences. As regards the human part of the population, the permanent effect of climate on the successive inhabitants of particular places is very marked in Greece. No one can visit the valley of Sparta without noticing the female beauty, which caused Homer to call Lacedæmon "the land of fair women." Nor have the Bœotians degenerated from their former boorishness. Almost the only place where I have found the Greek peasants rude was in that country, and, when I mentioned this to some friends at Athens, their remark immediately was, "Ah! Bœotia again!"

to know the first part of the way, we left Alipuchori, and passed along the fine valley under the eastern side of Mount Olytzika, which we had seen the preceding evening from the theatre. In the midst of this we found a number of beautiful fountains of pellucid water, which issue with a considerable stream from the rock, and bear the name of Kephalarisu; a village in their neighbourhood is called Milangus. The people spoke of them as one of the sources of the Acheron; but this is impossible, as the mountains of Suli intervene between. There can be little doubt from their position that they are the head waters of the Luro, the ancient Charadrus, which, like the Arta, finds its way southwards into the Ambracian Gulf. From this place to the village of Therike a rugged path leads through a very picturesque gorge, at each successive turn in which the mountains behind appeared to rise higher and higher. The inhabitants of Therike were squalid in appearance and dress, wearing dirty grey woollen leggings, and long tunics belted round the waist; their black and greasy scull-caps might once have been red, but if so, they retained no trace of their original colour. Being Christians they were of course unarmed, an inkhorn being the only implement that any of them carried in their belts; at the same time we heard even then of stores of arms that were being smuggled in from the coast, and concealed in various hiding-places, so that preparations were already being made for the coming insurrection. The village is prettily situated among upland meadows. Here we found a guide who undertook to direct us to Suli; and in his company we again descended the valley, and then turning to the right mounted a wooded ridge on the southern flank of Olytzika, the vastness of whose mass we were now able to appreciate from having seen it on three sides.

On the further slopes of this ridge is situated the little village of Toskis, where we halted for our midday meal. The natives, however, were most unwilling to receive us, fearing probably some forcible demands on their hospitality, and regarding all strangers as agents of the Turkish government. It was with difficulty that we obtained admission into one of the cottages, but at last they took us in, and supplied us with some provisions, for which we satisfied them by an ample remuneration. The view from this place was very striking. Before us lay a wide and undulating valley, bounded on either side by parallel chains of huge mountains, between which at the further end rose the grand peak of Crania. The northernmost of the two chains was that of Suli. We descended until we reached the foot of this, at a point about the middle of the chain, and then began to mount again, partly through woods, partly over open country. We had intended to pass the night at the little monastery of St. Saviour ("Αγιος Σωτήρ), which we understood to be in this neighbourhood; but our guide from Therike was now wholly at a loss, and, knowing nothing of the situation of the monastery, proposed, as a last resource, that we should make our way straight over the mountains to Suli. To this we acceded, but after having ascended a considerable distance we found ourselves about sunset on the open mountain side, without a path, in a very rough and shingly place, and with no very promising prospect either before or behind. The next difficulty arose from the *kharidji*, who declared that his horses could not go on; but at last we persuaded him, and dismounting, drove our animals up a rugged ravine or watercourse, along which they scrambled in an extraordinary manner. Just before the darkness had become complete we had the good fortune to discover a path leading upwards along a lateral ridge, which we followed until we gained the summit of the main chain: from the appearance of the vegetation, and the time occupied by our ascent, we apprehended that we must have reached an elevation of at least 3000 feet above the sea. We were now enveloped in clouds, and as we scrambled down we several times lost the path, and had to grope about until we found it again. At last, however, our difficulties were removed by our suddenly emerging into the most brilliant moonlight, which revealed to us an extensive and magnificent scene. Opposite to us rose a range of lofty mountains, and between these and the heights along which we were descending were seen at a lower elevation the peaks on which are built the castles of Suli, whilst in a grassy sequestered valley at our feet could be distinguished the dim outline of the ruined Kako Suli, with here and there the bright blaze of a shepherd's fire; and far beyond, and below, a winding line of silver appeared, where the Acheron was meandering in a plain towards the sea. Had it been daylight we should no doubt have seen Corfu, as these summits form a conspicuous object in every view from that island; as it was, the enchantment of the scene prevented us from desiring any additional features.

We had a letter to the Governor of one of the castles, but as this was still an hour's ride distant over a very bad road, and it was now between ten and eleven o'clock, we gave up the idea of reaching it, and determined to bivouac in the valley. The spot to which we descended was most romantic. All round us were the ruined homes of the Suliotes, retaining the same signs of desertion as when their inhabitants, after a brave defence, were driven out by Ali Pasha; no roof remained in any of them to afford shelter, and grass was growing on every hearth. We took up our position under a spreading tree, where some friendly shepherds joined us, and helped out our supper with goat's milk and cheese. A blazing fire was soon made, and it was a beautiful sight to watch the flames gradually rising, while in contrast with them the moon above—

> "rained about the leaf
> Twilights of airy silver."

As a great quantity of rain had fallen during the day, the circumstances were not wholly favourable to our open-air encampment; but as we had been in the saddle for fourteen hours, we had not to go far to seek repose, and wrapping ourselves up in our rugs on the ground were soon fast asleep.

On awaking in the morning we found the surrounding scenery more smiling, though not less grand, than we had conceived overnight. The summits of the lofty mountains were not generally visible from the valley, but merely the grassy slopes upon either side; trees and shrubs were growing luxuriantly in the midst of the ruined buildings in our neighbourhood, while, close below, a stream was rushing down on its way to join the Acheron. The gorge through which that river passes, and which from its narrowness and ruggedness bars all access to Suli from the side of the plain, lies far beneath, and is here excluded from view, owing to the elevation of this upland region. Towards the lower end of the vale, where it begins to descend towards this gorge, rose two tall peaks, one on either side, far famed for valorous deeds during the protracted resistance of the Suliotes. That on the left is called the "Ridge of Lightning," from the numerous storms which it attracts, and which have rendered futile all the attempts that have been made to construct a permanent habitation upon it; just beneath it on a lower elevation, called the hill of Kiafa, is built the Turkish fort. That on the right bears the name of Kiungi (Κιούγγι). The village of Kako Suli, which is the third point of chief historical interest, has often been compared in respect of this title to *Kakoilios*, the appellation of Troy in the 'Odyssey,' but the comparison is hardly a suitable one, as the latter term signifies "ill-fated Ilium," and was applied to the city after its destruction, whereas Kako Suli, or "Suli the Terrible," was so called from the terror which its inhabitants inspired into the neighbouring Mahometan tribes. It was at all times the head-quarters of the Suliote race, and occupied the most central position in the district.

The Suliotes were a body of Christian Albanians, who mustered some 1500 warriors when at the height of their fortunes, and were as independent of the Porte as the Mirdites are at the present day. They belonged to the tribe of the Tchamides, and in accordance with the usual strength and exclusiveness of clan feeling amongst the Albanians, they found their bitterest opponents in the members of another Toskish tribe, the Liapides, during their struggle with Ali. At the time of the Turkish conquest they had submitted, like all the surrounding tribes, and for several centuries they continued to pay the *haratch;* it was not until the commencement of the eighteenth century that they found themselves in a position to assert their independence, and to claim the right of carrying arms. Their possession of this privilege was at last officially recognized by the Pasha of Yanina somewhere about the year 1730. These armed Suliotes formed a military caste, and despised all labour as much as the proudest Mussulman. The soil in the richest part of their territory was cultivated by peasants who were of the Greek race, while the Albanian warriors ruled and protected them, as the ancient Spartans did their agricultural population. Their community soon became a refuge for other Christians of the tribe of the Tchamides, and when about the middle of the century, in consequence of this protection which they extended to others, they became involved in feuds with their Mussulman neighbours, they recruited their forces by admitting every daring and active young Christian of that tribe to serve in their ranks. If any of these volunteers distinguished himself by his courage, and was fortunate enough to gain booty as well as honour, he was admitted a member of the Suliote community, and allowed to marry a maiden of Suli. In this way they increased in numbers and in power.

As early as the year 1788, when Ali was appointed Pasha of Yanina, he proceeded to attack the Suliotes, in pursuance of his policy of centralising all power into his own hands, and destroying all the independent communities within his pashalik. He expected to subdue them the more easily, because they had become unpopular from their predatory forays into the territory of their neighbours, and were therefore likely to have to contend with him single-handed. But in this he was mistaken, for when he proceeded against them the other tribes became alarmed, and made common cause with them, so that they easily repulsed his first attacks. Four years afterwards, however, the struggle was renewed at the instigation of the Sultan of that period, Selim III., who discovered that the Suliotes were intriguing with Russian agents, and was led to regard their country as a nest of treason, as well as a nursery of brigandage. On this occasion Ali contrived to force the Kleisura or pass of the Acheron, and by means of his overpowering numbers gained temporary possession of Kako Suli. But he was unable to hold the position, and his losses in attempting to do so were so great that a large part of his forces deserted, and he was obliged to make a hasty retreat to Yanina.

These continued successes added greatly to the reputation of the Suliotes, but at the same time prosperity developed among them the two vices most deeply rooted in the Albanian character—pride and avarice, which were ultimately the cause of their destruction. By the former of these they were inspired with overweening confidence in their own power and importance, which led them to exercise their authority over the Christians in their territory with increased severity, and to plunder their Mussulman neighbours with greater rapacity. Owing to the other they were exposed to the machinations of Ali, who succeeded by means of bribes in detaching from the community several Suliote leaders, and enlisting them and their followers in his service. Among these was George Botzaris, a man of great influence, whom he afterwards employed against his countrymen. He did not, however, long survive his treachery, and his sudden death was attributed by some to a broken heart, by others to the effects of poison.

When hostilities were again recommenced, in 1799, Ali resorted to an entirely different plan of operations. Having previously failed in a direct attack on their stronghold he attempted a species of blockade, and by checking their forays, and occupying a number of strong positions on different sides of them, which he carefully fortified, gradually shut them up within narrow limits. At the end of two years, hunger began to be severely felt at Suli, and numbers of women and children were secretly removed to Parga, from whence they were conveyed to Corfu, which was then occupied by the Russians. In order to prevent further communication with that place the Pasha strengthened his posts on that side, and at the same time forced the Greek bishops to forbid the Christians in the neighbouring districts to lend any assistance to the Suliotes. In this state things continued until 1803, when the final struggle took place.

"The hero of Suli was a priest, named Samuel, who had assumed the strange cognomen of 'The Last Judgment.' It was said that he was an Albanian, from the northern part of the island of Andros; but he appears to have concealed his origin, for a hero in the East must be surrounded with a halo of mystery, though Samuel may have wished to erase from his memory everything connected with the past, in order to devote his soul to the contest with the Mussulmans, which he considered to be his chief duty on earth. He was an enthusiast in his mission, and, as he was doing the work of Christ, he cared little for the excommunication of servile Greek bishops. The Suliotes, who generally regarded every stranger with suspicion, received Samuel, when he first came among them as a mysterious guest, with respect and awe. At last, in the hour of peril, they elected him, though a priest and a stranger, to be their military chief. Religious fervour was the pervading impulse of his soul. His virtue as a man, his valour as a soldier, his prudence where the interest of the community was concerned, and his utter abnegation of every selfish object, caused him to be recognised by the soldiers of all the pharas as the common chief, without any formal election. His personal conduct remained unchanged by the rank accorded to him, and, except in the council and the field, he was still the simple priest. As he never assumed any superiority over the chiefs of the pharas, his influence excited no jealousy."[6]

It was through treachery that Ali ultimately obtained possession of the principal strongholds of the Suliotes. In the month of July, 1803, the clan of the Zervates, who occupied the castle of Kiafa, followed the example of George Botzaris, and abandoned that important

[6] Finlay's 'History of the Greek Revolution,' i. p. 60. The rest of this account of the Suliotes is mainly an abridgment of Mr. Finlay's narrative.

position to the enemy. Two months later the troops of Veli Pasha, Ali's second son, who was conducting the operations, gained possession of Kako Suli, in consequence of the treachery of Pylios Gusi, who, for a sum equal to about 300*l*., admitted 200 Mussulman Albanians into his house and barn during the night. After this the only points that remained in their power were the hill of Kiungi, and the village of Kiafa, near the foot of the hill of the same name. The former of these positions, where the Suliote magazines were placed, was entrusted to Samuel, and was defended by 300 families. The men guarded the accessible paths, posted behind low parapets of stone called meteris, and the women carried water and provisions to these entrenchments under the fire of the besiegers, who treated them as combatants. The number of women slain and wounded during the defence of Kiungi was consequently proportionably great. The little garrison dug holes in the ground under the shelter of rocks, and these holes, when roofed with pine-trees, thick layers of branches, and well-beaten earth, formed a tolerable protection from the feeble artillery which the Pasha had stationed, partly in their rear, on the summit of Ai Donato, and partly in their front on the Ridge of Lightning.

Before long, however, the Suliotes began to discover that the defence of their territory was hopeless, and were disposed to listen to the terms which Veli offered them. A capitulation was signed on the 12th of December, by which the Suliotes surrendered Kiungi and Kiafa, on condition that they might retire unharmed to Parga. Ali, however, who had determined from the first to violate the treaty, sent orders to place an ambuscade on the road and seize the Suliote chiefs, on whom he was thirsting to wreak his vengeance. But the Suliotes received timely warning of their danger, and by a rapid march and sudden change of route baffled his treacherous designs. Meanwhile the monk Samuel continued to occupy the fort of Kiungi, declaring that no infidel should ever employ ammunition entrusted to his care against Christians. When the Turks mounted the hill to occupy the place, he retired into the powder-magazine with a lighted match, and perished in the explosion, involving at the same time the foremost of those who were entering in his ruin.

But the tragedy was not even yet complete. Ali in his insatiable cruelty had determined on the destruction of the whole Suliote race, and now turned his hand against those who had concluded separate treaties with him. These he pursued in the most relentless manner, until almost all of them were either massacred or driven from the country. Some of the tales of this period are of the most heartrending description; but all bear witness to the intrepidity of the Suliote women, and their readiness to face death in the most terrible form rather than fall into the hands of their oppressors. One account relates how twenty-two women of one village, being driven to despair, cast themselves over a precipice, down which they had already flung their children. But the most famous story, which has been embodied in a Greek ballad, is that of Despo, the wife of a Suliote called George Botsis, who, when her dwelling was surrounded by the Mussulmans, followed the example of the monk, and set a light to a chest full of cartridges, by the explosion of which she and her family were blown to atoms.[7] The following is the ballad :—

---

[7] The story is given in Fauriel, 'Chants Populaires,' i. p. 279.

"A loud discharge is heard afar, the crack of many rifles:
Say, is it for a marriage feast, or joyous celebration?
It is not for a marriage feast, or joyous celebration;
'Tis Despo and her family, who gallantly are fighting:
Within the tower of Demulá th' Albanians besiege them.
'Thou wife of George, lay down thine arms; deem not thou art in Suli;
Lo! here thou art the Pasha's slave, thou art the Albanians' vassal.'
'And what if Suli is subdued, and Turks possess Kiafa?
No master of Liápid race has Despo served, or will serve.'
She seized a firebrand in her hand, she called her daughters round her;
'We'll never live as Turkish slaves; one last embrace, my children!'
A thousand cartridges were there, she threw a light among them;
The cartridges flared up on high, and in the blaze they perished."[8]

On leaving the ruins of Kako Suli we took one of our shepherd friends as a guide, and wound down the valley towards the gorge of the Acheron. It was a rugged track, and led through a most desolate region; accordingly, great was our surprise when on passing the foot of the hill of Kiafa, beneath the Turkish castle, we met a picturesque procession, a bridal party in their best array, singing gaily to the music of a violin and tambourine. On such occasions in Albania it is possible to estimate the money value, as well as the charms of the bride, as she carries all her fortune on her person in the shape of ornaments, principally pieces of money. In this instance the fair lady was wealthy, for her headdress and other parts of her toilette were formed of strings of coins, silver and gold, amounting to a truly handsome dower. However, the gentleman is usually required to pay down a certain sum to the relations of his future spouse, as is testified by the following little Albanian love poem, in which the lady is represented as

---

[8] Passow, 'Carm. Pop.,' No. 214.

regarding the amount of the money as a guarantee of her lover's affection for her. The gentleman begins—

"Lovely little mistress mine,
My citron thou, my orange fine;
A lordly lover should be thine!"

The lady answers—

"No lordly lover, dear, for me;
Only my fond one's spouse I'll be,
Who three hundred crowns has paid,
All to win his tender maid."[9]

Beyond this point we commenced a steep descent through scenery of a magnificent description. The crags rise steeply to an immense height above, and the path, which runs along the side of the hills, hangs far above the valley, while the glens which are furrowed in the mountains are everywhere filled with trees. We were forced to send our baggage-horses round by another route, though not without some misgivings on account of the wild character of the neighbouring population; but every one agreed that it was impossible for them to make their way through the gorge. At length we reached the Acheron, whose white waters are here confined within a rocky bed, and roar along through chasms and clefts which they have worn away in the course of ages. The sides of this passage are, for the most part, richly clothed with foliage, trees and shrubs clinging to every available point, whilst among so much luxuriance the light-grey rocks peep out in the most enchanting manner: above this, again, the mountain summits tower at a great elevation on either side. The gorge continues between two and three miles, and is in its way unrivalled

[9] Hahn, 'Albanesische Studien,' ii. p. 128.

even in Greece. We proceeded for some little distance by the side of the stream, until we came to a small bridge composed of branches, hanging high above the water, by which we crossed. Our saddle-horses, however, had to be taken across at a point some way higher up the stream, where, according to our dragoman, they were made to jump from rock to rock, with the continual risk of falling into the swift current, until at last they reached the other side. We then left the river and ascended among the woods, through the openings in which there were lovely views, the castle of Suli standing out in bold relief as we gazed back up the valley. The finest point of all is at a turn of the road near the middle of the pass, where you can see the Acheron running between precipitous rocks directly below you at a depth of 500 feet, and hear the dull roar of the surging waters softened by the distance. Nothing intervenes between you and the stream but a few trees, which have fastened their roots in the fissures of the cliffs. A little further on, where the valley begins to open out, there is a rough path, by which we scrambled down to the river, leaving our horses to follow us as best they could. While I was sketching at the mouth of the gorge, my companions bathed, and found the water excessively cold. This was mainly owing to several large springs, called the Great Source, which here issue from the rock, and by their pure blue water flowing through the white river recall the meeting of the Rhone and the Arve at Geneva.

Some way further down we forded the Acheron. This was a matter of some little difficulty, for its stream is deeper than that of most of these mountain rivers, and the current was so swift that our horses could hardly keep their footing. When we reached the further bank, and found our baggage animals safely awaiting us, we

GORGE OF THE ACHERON.

felt glad that we had sent them round. Below this point the ground is low and marshy, and the river forms a considerable lake in the winter, which in ancient times was the Palus Acherusia. At this season, however, there was no water visible, but we soon discovered where it had been; for, having dismissed our Suliote guide, as we expected to be able to find our way to Parga without difficulty, we were soon floundering about in a quagmire, from which we had some work to extricate ourselves. On the northern side of this is the village of Glyky, the name of which seems to be a perpetuation of that of Glykys Limen, the Sweet or Fresh Haven, as the port was called, where the Acheron enters the sea a few miles lower down. In the church of this place Dr. Wordsworth found remains of the column of a temple, and there is reason to believe that this is the site of the ancient oracle of the dead from which Herodotus tells us that the Greeks of his time used to seek responses.[10] Pausanias expresses his opinion that Homer derived the idea of his 'Inferno' from this spot, and adopted the names of the rivers of this part of Thesprotia.[11] Lofty rocks, as well as rivers and a marsh, certainly entered into the Greek conception of these regions; thus Aristophanes speaks of "the cliff of Acheron dripping with blood."[12] When the Greek conception of the nether world was imported into Italy and localized there, its impressiveness was lost: nothing can be less appropriate to it than the Palus Acherusia of the Romans, the modern Lago Fusaro near Naples, which is a mere lagoon. In like manner the Styx was associated in the mind of the Greeks with the magnificent waterfall, which descends 500 feet over a stupendous cliff in the wildest part of

---

[10] Herod., v. 92.   [11] Pausan., i. xvii. 5.   [12] Ar. Ran., 471.

Arcadia, and was only partially connected with the lower world; whereas the Romans conceived of it only as a sluggish stream, wandering in nine channels round the infernal regions.[13] I know nothing which so well illustrates the disposition of the Greeks to interpret their profoundest ideas by the help of grand natural objects, as these two rivers, the Styx and the Acheron.

The path now bore away to the westward, along the bare slopes which here rise above the low ground, in the midst of which appeared the village of Turcopali. From that point it led over a wild hilly country in the direction of the sea, thick-set with bushes of prickly palluria, and afterwards for some distance above the rugged coast. At length the islands of Antipaxo and Paxo, and the more distant southern promontory of Corfu, appeared in view, announcing that we were approaching the end of our journey. After five hours' and a half riding from the time we left the Acheron, we found ourselves close to the beach at Parga, on which we unexpectedly emerged, after passing through a grove of gnarled olive-trees, with which the hills on the land side of the town are covered. The situation of the place is most picturesque. The coast at this point faces south, and a small natural harbour is formed by two or three rocky islands which lie in front of it. Upon a cliff at the western horn of this bay stood the old town, now marked by ruined houses and a Turkish citadel. Eastwards rises the smaller eminence of St. Athanasio. Outside the fortress stands the new town, extending down to the beach, with one white minaret, denoting it to be an Osmanli possession.

The first thing that we did on our arrival was to

---

[13] *E.g.*, compare Hesiod, 'Theog.,' 785-792, with Virgil, 'Georg.,' iv. 480.

charter a small vessel to convey us to Corfu. It was late at night, however, before we started, and we amused ourselves, as we sat on the shore, by watching the brilliant moonlight on the cliffs and the sea. For myself, I was dreaming over some scene in 'Massaniello,' when our dragoman returned from purchasing some provisions in the bazaars, and accosted us reproachfully with the words,—"The people here say that you betrayed them." We received the announcement with the stolidity of true Britons; but there was something unpleasant in being reminded on the spot of a transaction, which is certainly a blot on our national history. The cession of Parga caused a great outcry in Europe at the time, and the romantic and pathetic circumstances which accompanied it have given it a permanent interest. As some of my readers may not remember these, I will here introduce Sir A. Alison's account of the event :—

"The town of Parga, on the sea-coast of the main-land, opposite to the Ionian Islands, the last remnant of the once great territorial possessions of the Venetian Republic on the coast of Albania, had long been considered as a dependence of the state of which they had come to form a part; and in the interval between its cession to France by the Treaty of Tilsit, in 1807, and its transference to Great Britain by that of 1814, it had contained a French garrison, and its inhabitants had begun to taste the blessings of powerful Christian protection. The Treaty of 1815, however, unfortunately made no mention of Parga, but, on the contrary, stipulated an *entire* surrender of the mainland of Turkey to the Porte. In consequence of this circumstance, the Government of Constantinople demanded the cession of Parga as part of the main-land; and in this they were zealously seconded by Ali Pasha, within whose territory it was situated, and who was extremely desirous of getting its industrious and thriving citizens within his rapacious grasp. On the other hand, the inhabitants of Parga, justly apprehensive of the consequences of being ceded to that dreaded satrap, solicited and obtained a British garrison, which in 1814 took possession of it, and effectually preserved its inhabitants from Mussulman rapine and rapacity. The inhabitants joyfully took the oath of allegiance to the English Crown. Thenceforward they regarded themselves as perfectly secure under the ægis of the victorious British flag.

"When it was rumoured, after the Treaty of 1815, that Parga was to be ceded to the Turks, the inhabitants testified the utmost alarm, and made an urgent application to the British officer in command of the garrison, who, by order of Sir Thomas Maitland, the governor of the Ionian Islands, returned an answer, in which he pledged himself that the place should not be yielded up till the property of those who might choose to emigrate should be paid for, and they themselves be transported to the Ionian Islands. An estimate was then made out of the property of the inhabitants, which was found to amount in value to nearly 500,000*l.*, and the inhabitants were individually brought up before the governor and interrogated whether they would remain or emigrate; but they unanimously returned for answer that 'they were resolved to abandon their country rather than stay in it with dishonour, and that they would disinter and carry with them the bones of their forefathers.' Commissioners had been appointed to fix the amount of the compensation which was to be awarded by the Turkish government to such of the inhabitants of Parga as chose to emigrate; but they, as might have been expected, differed widely as to its amount, and in the end not more than a third of the real value was awarded. Meanwhile Ali Pasha, little accustomed to have his demands thwarted, and impatient of delay, repeatedly threatened to assault the town and re-unite it to his pashalic, without paying one farthing of the stipulated indemnity. At length, in June, 1819, the compensation was fixed at 142,425*l.*, and Sir Frederick Adam gave notice to the inhabitants that he was ready to provide for their embarkation.

"The scene which ensued was of the most heart-rending description, and forcibly recalled the corresponding events in ancient times of which the genius of antiquity has left such moving pictures. As soon as the notice was given, every family marched solemnly out of its dwelling without tears or lamentation; and the men, preceded by their priests and followed by their sons, proceeded to the sepulchres of their fathers, and silently unearthed and collected their remains, which they put upon a huge pile of wood which they had previously collected in front of one of their churches. They then took their arms in their hands, and, setting fire to the pile, stood motionless and silent around it till the whole was consumed. During this melancholy ceremony, some of Ali's troops, impatient for possession, approached the gates of the town, upon which a deputation of the citizens was sent to inform the English governor that if a single infidel was admitted before the remains of their ancestors were secured from profanation and themselves with their families safely embarked, they would instantly put to death their wives and children and die with their arms in their hands, after having taken a bloody revenge on those who had bought and sold their country. The remonstrance was successful; the march of the Mussulmans was arrested, the pile burnt out, and the people embarked in silence with their wives and children."[14]

The event here described has been celebrated in a spirited Greek ballad, with which I conclude my narrative:—

" There came three birds from Prevyza, and flew across to Parga,
One turned his face to foreign lands, and one to Saint Jannaki ;
The third, whose plumes were raven-black, uttered his lamentation:
' Parga, the Turks have hemmed thee in, the Turkish bands surround thee ;
'Tis not for war that they are come, by treachery they've won thee ;
Not by the might of all his hosts has the Vizir subdued thee ;
Like trembling hares the Turks have fled before the guns of Parga,
And the Liápids had no heart to meet them in the conflict.
Thy warriors were like beasts of prey, thy women bold as heroes ;
Bullets to them were e'en as bread, powder the meat they fed on ;
But 'twas for silver Christ was sold, and thou art sold for silver.'
' Ye mothers, bear your children hence, ye priests the holy relics,
And leave, ye warriors, your arms; ye warriors, leave your muskets ;
Dig up the graves, dig wide and deep through all your burial-places,
Exhume the consecrated bones of your heroic fathers ;
To Turks they never bowed the knee ; no Turk shall trample on them.' "[15]

---

[14] Alison's 'History of Europe,' 1815-1852, iii. pp. 86-88.
[15] Passow, 'Pop. Carm.', No. 223.

CHAPTER XXVIII.

THE ROMAIC BALLADS.

Ballads of Modern Greece — Their Literary Value — Collections of them — Absence of Antiquity — Their Romantic Origin — Classification of them — Leading Characteristics — Homeric Features — Idyllic Pieces — Story of the Bridge of Arta — "The Abduction"— Dirges or Myriologues — Poems relating to Charon — Love Poems — Elaborate Rhyming Poems — Distichs — The "Political" Verse — Italian Origin of the Rhyme — Mode of Singing them — The Ballads an Index of the National Character — Songs of Farewell — Pathos — Absence of Humour — Sympathy with external Nature — Greek Nursery Rhyme.

THE ballad poetry of modern Greece is a subject to which, unfortunately, but little attention has been paid in England. At one time, indeed, it seemed probable that it would have been otherwise. The sympathy which was aroused by the Greek War of Independence, and the few specimens which were translated by Lord Byron, seemed likely to divert the taste of our literary men into this channel; and accordingly it was shortly after the conclusion of that war that the only contribution worth mentioning, which any Englishman has made to that branch of literature, was formed, viz., the large collection of Cretan ballads and distichs published by Mr. Pashley in his 'Travels in Crete.' Since that time the interest in the subject appears to have flagged, and it is only in rare instances that scholars have been found to pay attention to it. For several reasons this is much to be regretted; for not only are these poems of considerable merit in themselves, and illustrative of the character of a very remarkable people, but the language in which they are expressed, and which is found in a literary form in them alone (for the Neo-Hellenic, in which the publications of modern Athens and Corfu are written, is to all intents and purposes a different language), is highly valuable for philological purposes. In it we have a specimen of a language exposed to similar influences, and undergoing similar changes, to those which have befallen the modern languages derived from Latin, while it has been only in a very slight degree affected by those languages, so that it affords excellent opportunities for independent comparison. Again, notwithstanding the unreasonableness of most of the attempts that have been made to trace a connection between the modern tongue and particular dialects of the ancient language, there is not a little to be gained from it in the way of illustration of classical forms, *e.g.*, of the use of the digamma,[1] and especially of various peculiarities of usage in the Greek of the New Testament. And to the student of Byzantine and later Greek history there is much to be learned with regard to the changes that passed over the country, from a careful examination of the classes of words which have been imported into the language from Roman, Slavonic, Turkish, and a variety of other sources.[2]

---

[1] Perhaps the most remarkable instance is αὐγό (pronounced *avgo*), the modern Greek for "egg." This is the original Indo-European form of the word, the primitive *a* of which was changed, both in Latin and Greek, into *o*, while the former language dropped the *g*, and formed *ovum*, and the latter cast out the *v*, or digamma, thus making ὤιον (Æolic) and ᾠόν. The modern Greek has preserved all the earliest elements. Our "egg" is a corruption of the same word. See Curtius, 'Grundzüge der Griechischen Etymologie,' p. 351. This author fully recognises the philological importance of modern Greek.

[2] Thus Latin, from having been the language of the court, the public offices, and the army, from Constantine to Justinian, is now mainly represented by words derived from those sources. For instance, through the usage of the court are derived ῥήγας, "a chieftain," from *rex*, which term was mainly applied by the Byzantines to Western potentates; βέργα, "a rod," from *virga*, introduced by the use of the fasces. From the public offices come βούλλα, "a seal" (*bulla*); σεκρετάριος, "a secretary"

Meanwhile the work which was neglected in England has been accomplished by Frenchmen and Germans, and by the modern Greeks themselves; and the result is, that by degrees a very large collection has been formed, more complete in all probability than any set of ballads of any other people, and is receiving constant additions. The poems are of very various merit, ranging from mere doggrel up to highly poetical pieces; but the collectors have acted wisely in excluding none because they appeared commonplace or second-rate, for, in consequence of this, the collection embraces every kind of subject, and illustrates the life of the nation, its ideas, superstitions, and inner feelings, under very varied circumstances, in a manner to which it would be hard to find a parallel. The collector who was first in the field was fortunately a man eminently qualified for the task. This was M. Fauriel, who is best known for his works on the Provençal language and on Dante, and who was at once a philologist, a man of taste, and an excellent critic, as well as being well acquainted with the Greeks, and sympathising with their cause. The two volumes which he published shortly after the commencement of the War of Independence, under the title of 'Chants populaires de la Grèce,' though they have now been superseded as a collection, still retain a permanent value from the excellent sketch

(*secretarius*). From the army, βίγλα, "a watch" (*vigilia*); ἅρματα, "arms" (*arma*); παγανία, "the militia" (*paganus*); *i.e.*, the soldiery of the rural districts as opposed to the standing army. On the other hand, many of the nautical terms are derived from Italian, having, no doubt, been introduced by the Venetians. Such are κουμπασάρω, "to observe the compass" (*compassare*), τιμόνι, "a rudder" (*timone*); σιγουράντσα, the name for one of the cables (*sicuranza*). Curiously enough the reverse process took place in ancient times, when the nautical terms passed from Greece to Italy,—*gubernare*, *ancora*, *prora*, *aplustre*, and similar words in Latin, having been originally Greek. Many other points of interest may be worked out in the same manner.

of the literature contained in the Introduction, and the critical notices dispersed throughout the work.[3] Amongst the many Germans who have subsequently laboured in the same field, the name most deserving of mention is that of the distinguished scholar and traveller Professor Ulrichs. Though he was prevented by death from editing with his own hand the numerous ballads which he had discovered, yet the work fell into good hands when it was entrusted to his son-in-law Arnold Passow, who published these in 1860 together with all the other ballads which had been found up to that time—amounting to between six and seven hundred in all, besides distichs, under the name of 'Popularia Carmina Græciæ recentioris.' In this edition the different versions of the same poem, where more than one exists, have been carefully compared, and a Latin introduction and a glossary of rare words have been added: and though there is something comical at first sight in such headings as *Carmina Clephtica incerti ævi*, and in the apparatus of various readings appended to each ballad, yet too high praise cannot be given to the careful way in which the work has been done, especially in respect of the difficult question of orthography. In this it has been attempted to follow as far as possible the pronunciation of the people,—a point of great importance, in order to discover the true forms of the language, and to note dialectical peculiarities. The poems may be roughly classified under the following heads: (1.) Those that turn on historical incidents, and on the exploits and adventures of the Clefts. (2.) Domestic songs, including nursery rhymes, songs composed for certain festivals and seasons of the year, and "farewells," or poems recited by persons leaving their homes. (3.) Dirges, and other

[3] I am indebted to Fauriel's book for much of what has been introduced in the following pages.

songs relating to death and the state of the dead, which are so numerous and remarkable as to deserve to be classed by themselves. (4.) Pastoral and imaginative pieces. (5.) Love songs, and those which accompany the numerous ceremonies connected with Greek marriages.[4]

Out of this large collection there are but half a dozen[5] which have any pretence to considerable antiquity, and two of these are shown to be more modern than the date usually assigned to them, by Turkish words or words of later introduction which they contain. The other four may very possibly be ancient, viz., one referring to the taking of Adrianople by the Turks in 1360, which may be contemporary with that event, and three others relating to circumstances connected with the taking of Constantinople: it is no argument against their antiquity that the Greek in which they are composed is the same in form with the Romaic of the present day, for there is evidence to show that the popular dialect, which had become a separate language from that in use with the Byzantine writers as early as the First Crusade, had by the middle of the fifteenth century assumed all the peculiarities of form and syntax which we find now to belong to it. But of the other ballads hardly one can be placed further back than 150 years from the present day. At the same time there is no reason to suppose that the art of composing these, or their wide distribution among the people, dates from recent times. On the contrary, there are pieces embedded in some of the

[4] The modern name for all these songs and ballads is τραγοῦδι. This word is found in the mediæval Byzantine romances in the sense of a poetical "complaint" of an amatory character; thus it is derived from the original "tragedy," on account of the emotional element it contains.—Gidel, 'Études sur la Littérature Grecque Moderne,' p. 167.

[5] Nos. 193 foll., in Passow's collection.

poems that have come down to us from the Byzantine period, which are in all respects analogous to some of the modern songs; and Fauriel has suggested reasons for thinking it not impossible that this style of composition may have descended, more or less directly, from the popular poetry which is known to have been in use in Greece at certain seasons of the year, or among the members of certain trades and professions in classical times.[6]

Nothing can well be more romantic than the origin of many of these ballads. Some are composed and sung by shepherds on the mountains: some by sailors in the course of their voyages; while others, in all probability, were the work of the Clefts themselves. Again, the dirges, or *myriologues*, as they are called, are in some cases extempore effusions recited by women under the influence of violent emotion over the bier of some dead relation. But by far the greatest number owe their origin to a class of men who here, as among the Montenegrins and the Greeks of the heroic age, are the great depositories of the national poetry, the blind wandering minstrels. Throughout the country it is an established custom that the blind man should obtain his livelihood by composing and reciting ballads, of which some of them are said to know an incredible number by heart; and in his character of rhapsodist he passes from village to village, selecting especially for the time of his visit some local festival, when a large audience will be gathered together, from whom he never fails to obtain an enthusiastic hearing and bountiful contributions. He is at the same time the author of the music to which each ballad is

---

[6] 'Chants Populaires,' vol. i., Discours Préliminaire, pp. c. foll.

sung, and accompanies himself during his recitation on a small guitar. Many of these songs have come to be used for dancing to, and may frequently be heard on such occasions sung in chorus by the whole body of dancers. But in all cases, it should be observed, and whatever may be the source from which they are derived, they are the offspring of spontaneous, untutored genius, and their authors are persons who can neither read nor write, and come to the task with no special preparation; and whose object is not to win fame, for their names are in no case attached to their compositions, but to give vent to a feeling or an idea which they desire to express in words. Many of these peculiarities in the origin and character of the ballads are so well illustrated by a passage of Ulrichs' 'Travels in Greece,' where he is giving an account of the village of Arachova on the side of Parnassus, that I cannot forbear to introduce it here:—

"Arachova is a place where the life of the country-people of Greece presents its fairest characteristics. Independent and prosperous, and blessed with vigorous health, its inhabitants pass the summer months partly in tilling their fields and pasturing their flocks in the breezy upland valleys, partly in cultivating their vineyards on the declivities of Parnassus. During the winter, when the snow-storms and the cold northerly blasts sweep around them, they return each of them to his comfortable home. From their numerous festive meetings, and the joyous public celebration of marriages and other festivals—generally of a domestic character—but, above all, from their custom, handed down to them from primitive times, of dancing in large bands and accompanying the figure with song, poetry becomes indispensable to them. Innumerable poetical productions shoot up like plants in spring, and disappear again, while here and there a song of remarkable merit obtains a more durable existence and wider circulation without any one giving himself the trouble to inquire its composer's name. When a bright thought or felicitous expression strikes a person's mind, he adapts it to this in singing; and thus a poem grows in the mouth of the people and arrives at a complete and permanent form, from which at the time when it was first improvised it may have been many degrees removed.[7] As of the Homeric poems, so also of those of modern Greece, it may be said that an entire people combined to form them and mint them into Popular Songs in the truest sense of the word. All that the latter want is a great theme, which might make combination possible. When on festival days I see a company, and seated in the midst of them an old musician singing to a simple guitar his lays, in which the choir of dancers join, I cannot help recalling Homer's words—

> 'Then took the bard divine the hollow lyre,
> And of the sprightly dance and music sweet
> Stirred in their breasts a very warm desire.'[8]

Nor must we be misled by the monotonous and discordant melodies of the modern Greeks as they appear to our ear, so as to draw an unfavourable conclusion with regard to the words of what is sung. The Greeks are fond of their tunes; and the contents of a song, whatever the music may be, will often throw them into the deepest emotion, so that neither the singer nor his hearers can restrain their tears."[9]

In considering the style and characteristics of the poems it may, perhaps, be most convenient to notice first those which, from their narrative character and marked simplicity of composition, possess in the highest degree the peculiarities of ballad poetry: next, those which approach more nearly to idylls, being imaginative pieces on ideal subjects, somewhat more elaborately treated: and lastly, the love poems, serenades and other songs, some of native growth, others evidently modified by external influences, in which a higher culture and greater skill is traceable. It will be found for the most part that the first of these divisions will correspond to those composed in the mountains, while those comprised

---

[7] See what has been said above (p. 95), on the formation of the ballad of "The Spectre." The remarks of Ulrichs which follow are well worthy of consideration, as there is nothing else in modern literature which appears so well fitted to elucidate some of the most difficult points in the question of the authorship of the Homeric poems as these ballads.

[8] Hom. 'Od.' xxiii., Worsley's translation, stanza 19.

[9] Ulrichs' 'Reisen und Forschungen in Griechenland,' i. 130, 131.

in the two latter mainly owe their origin to the cities, the islands, and the coasts of the Ægean. But since there are many characteristics which are common to almost all the poems, it may be well in speaking of the first class to mention once for all these general features.

As the Clefts were the national and local heroes of the modern Greeks at the most stirring period of their history, the simple narrative ballads, as might be expected, treat mainly of their deeds of prowess. A considerable number, however, are devoted to the commanders in the War of Independence and to incidents in that struggle, together with some of the most marked events in the history of Ali Pasha and the wars of the Suliotes. From the uniformity of the subjects of which they treat they are in some respects the least attractive part of the whole collection; and the large number of Turkish words that occur in them, and the roughness of the dialects in which they are composed, render them the most difficult to read: but the rude and masculine vigour of their style, their bold straightforwardness of expression and frequent abruptness of transition, harmonize well with the deeds which they celebrate, and bring before the reader in a striking way what they are intended to describe. All these peculiarities are well illustrated in the ballad of 'Olympus,' which has been already given in p. 51. One noticeable feature, both in these and in the other poems, is their conciseness: one occurrence, or idea, or touch of feeling, is selected, and this is seldom treated at any great length, but is put forward in a series of bold touches which, without any apparent attempt at harmonious arrangement, succeed in leaving a distinct impression on the mind. As has been already suggested, there is hardly the least trace of rules of art, so that any signs of such an influence must be referred partly to the good taste of the people at large, and partly to the feeling of the individual composer: at the same time there are a variety of forms which appear traditional and are constantly found to recur. Thus it is not uncommon to find the subject introduced by a sort of prologue, which either suggests the tone of feeling which is about to pervade the poem,[10] or introduces an independent narrator, who is to relate the story. Thus, when the death of a hero is to be recounted, the tidings are frequently represented as being brought by one or more birds, which take up their position on some conspicuous cliffs, and from thence utter their tale of woe. Again, in them, as in the ballad poetry of other nations, fixed epithets regularly recur, applied to particular objects and persons, and the practice of repeating certain forms of expression, which has been called "Epic common-place," is of constant occurrence. Just as in Homer we are accustomed to stereotyped phrases for the introduction of a speech, for embarking and disembarking when on a voyage, for cooking and carving at meals, and for numerous other occasions, so in the ballads there are regular forms to express transition in the narrative, or the occurrence of an interval of time, or passage from one place to another, and for other uses corresponding even more closely to those just mentioned. In both of them this custom is intended to be of use as a help to the reciter's memory, and it is always grateful to an untutored audience, who are not fastidious enough to require constant variety, and are fond of anything resembling a burden or refrain. But its use is far more frequent in these ballads than in Homer; indeed the

---

[10] See, for instance, that introduced below, p. 327.

recurrence of certain expressions, and that too not always relating to the most ordinary circumstances, in great numbers of the poems which are found dispersed over a very wide area, seems to imply that they date from a considerable antiquity, and have become the common property of the nation through gradual transmission, and in particular by the agency of the wandering bards. They are, in fact, the constant element in a literature which is for ever shifting, and perishing to be renewed again, having been selected by a continual process of elimination, and tested by repeated use.

Another point of similarity between these ballads and the Homeric poems, is the way in which in both the subject is dramatized. It has been remarked of Homer that he withdraws himself from view in his character of narrator, whenever that is possible, and describes his personages either by making them think aloud, or by allowing the accessory circumstances to bring out their characteristics, or by some other indirect method. The same thing, only in a still greater degree, is true of the ballads. In them descriptions are almost, if not wholly, wanting, and it is the rule rather than the exception for the narrative form, at least in some portions of the poem, to be so merged in the dramatic as to be completely lost. To the reader the effect of this is to produce a sense of abruptness and roughness of composition ; and here and there in dialogues the connection of ideas is sufficiently obscure to lead one to think that some explanation must have been required by means of gesture or tone of voice; but in recitation the impressiveness of the poem must have been heightened by this means, and the scene or event it described have been brought more vividly before the audience.[11] If, in addition to these peculiarities, we notice the rapidity of movement in these pieces, the manner in which the subject is concentrated, and the passionate warmth and tenderness of the feelings expressed, we shall appreciate the truth of the remark, that one great secret of their charm is the simple and unstudied way in which the epic, dramatic, and lyrical elements are combined in their composition.[12]

After what has been said, in connection with Mount Olympus, of the history of the Clefts and Armatoles, it is hardly necessary to suggest to the reader the close resemblance between the persons celebrated in these poems and the Homeric heroes, or the chivalrous knights of the Middle Ages, though they are distinguished from them to some extent by a more savage nature, and by the absence of many civilising influences. In all of them is found the same independence, the same sense of personal honour, the same undaunted bravery and passionate enthusiasm, the same spirit of self-assertion. In the descriptions of the accompaniments and surroundings of the modern Greek heroes, however, there is, as Fauriel remarks, a tone of exaggeration which is Oriental rather than Hellenic, and which is more or less characteristic of all the poems. Thus the Clefts' horses are described as shod with silver, their persons are resplendent with gold ornaments, and pearls and diamonds are scattered about with a lavish profuseness only found in Eastern tales. And in other respects, while in Homer the objects in

---

[11] In some of the ballads it even happens that, though the narrative is commenced in the first or second person, it is continued after a time in the third; this is the case with the original of the ballad of "The Spectre."

[12] Thiersch, 'Ueber die Neugriechische Poesie,' p. 32.

ordinary use are described with a faithful minuteness, which enables us now to realise many of the details of the life of the period, the ballads aim at idealising everything commonplace, and investing it with imaginary splendour. Another point in which they differ in a marked way from epic verse arises from the nature of the metre, which owing to its length is usually capable of expressing an idea fully in one line, so that the sense is but rarely carried on from one verse to another. And as the metre naturally divides itself into two hemistichs, the latter of which is the weaker of the two, it possesses some of the characteristics of elegiac verse, the latter half being frequently a repetition or amplification of the meaning of the former, by which means a sort of balance is produced, and greater stress is laid on a statement or idea when that effect is required. This process is facilitated by the extreme simplicity of the construction of the sentences, which show that the language has been formed by a people in a very primitive state of civilisation, co-ordinate sentences being constantly found where subordinate ones might have been expected.

We next come to the most interesting part of the whole collection, viz., those poems which may be regarded as idylls. If an idyll may be defined as a short, complete, and elaborate descriptive poem, they very properly deserve the name; and the subjects of which they treat correspond in great measure to those which occur in other collections of idyllic poetry, being either pastoral pieces, or laments and dirges, or romantic and pathetic stories, or the development of some superstition or mythological fancy. From this it will be seen that they are sometimes purely imaginative, and at others based on some real or traditional circumstance, ideally treated. In them the sentiment is more elevated, the diction more refined, and the theme more varied, than in those of which we have hitherto been speaking. They also reflect from very different sides, and exhibit in a variety of aspects, the character of the whole people from whom they originate, in all its brilliancy and versatility. And there is an additional interest attaching to them, if we regard them as affording us an analogy by which to judge of the origin of Greek bucolic poetry in classical times. According to the only ancient authorities who have touched upon the subject, the pastoral poems which attained to such perfection of treatment in the hands of Theocritus were originally derived from the rude songs of shepherds in Sicily and on the continent of Greece,[13] in which case they must have come to light in the midst of the same associations and external influences—the same scenery, and sky, and climate, and mode of life—and must in all probability have been developed and refined by the same process, as those of which we are now speaking.

In the pastoral pieces the central figure is not unfrequently a Wallach shepherd, the nomad families of that race being in many parts of Greece, as we have seen on Mount Olympus and elsewhere, amongst the most hardy of the mountaineers. They treat of the shepherds' distresses and loves, of the incursions of robbers by whom their flocks are carried off, of wrestling bouts with Charon, in which the latter always comes off victorious, and similar subjects. Here and there allegory is introduced into them, though it is not very commonly found in the poems at large, and appears to have been suggested by deep emotion, as it is usually employed to express the distressed condition of the people. Thus the aged Cleft

---

[13] *See* Ulrichs' 'Reisen in Griechenland,' p. 140, *note* 6.

who is being hunted down by the Turks on the mountains, and is thus regarded as a type of his countrymen generally, is represented as the old stag or old eagle, which now that its speed or strength of wing has departed from it is being surrounded by its pursuers and their dogs, and has no longer any chance of escape.[14] Another ballad tells the story of the nightingale, how she attempts to build her nest in a number of places near the haunts of men—in a garden, on the poop of a vessel, and by the sea shore, and when driven away from all of them, she is seized with despondency, and flies to a solitary tree near the window of a fair princess, and there pours forth her song of love. The princess hears her, and envies her happiness and her sweet strains, on which the bird contrasts the lady's life of ease and safety with her own melancholy fortune.[15] And in a little poem brought from Cyprus, two doves are represented as talking in their nest, and the male chides his mate for her joy at her newly-laid eggs, by reminding her how birds of prey and rude hands had deprived them of their former brood, and how little hope there was of their remaining unmolested.[16] Closely connected with these, again, are the songs relating to some object of popular superstition; such as the genii which haunt the woods and ravines, the power of magic, the Nereids, who are conceived of somewhat differently from what they were in ancient times, and other beings, of whom more will be said presently, when we come to speak of the mythology of the modern Greeks. These beliefs, from the deep impression they have made on the mind of the people, have given rise to some of their finest compositions.

Among the imaginative pieces which turn on some

[14] Passow, Nos. 501, 502. [15] Ibid., No. 449. [16] Ibid., No. 403.

romantic incident, whether fictitious or otherwise, those that relate to the Bridge of Arta deserve to be first mentioned. The story seems to have been suggested by the vibration of the arches of the bridge, and runs as follows. A number of masons had long been engaged in building the bridge of Arta, but what they constructed during the day fell down during the night. When they were sorely perplexed about this, a voice was heard to proceed from the bridge, saying that the work could not be accomplished except at the sacrifice of the life of the master-mason's beautiful wife. On hearing this, the master-mason was at first overwhelmed with grief, but ultimately the desire of carrying out his design prevailed, and he sent a message desiring her to come in her richest dress and her jewels. When she arrived, and asked the reason of his summons, he replies that his wedding-ring had dropped into the hollow of one of the piers they were constructing, and that she was the fitting person to regain it for him. She at once consents to be let down; but as soon as she has reached the bottom, they commence heaping lime and mortar upon her, and her husband himself casts down the first stone. As she dies her voice is heard to exclaim, that like as her heart quivers at that moment, so the arches of the bridge shall quiver for ever and ever.[17]

To the same class may be referred a number of poems highly characteristic of the condition of the people, those, for instance, which describe the recognition of husbands by faithful wives after many years' absence. In these the unexpected visitor, whose appearance has been changed by time and other causes, on returning to his native vil-

---

[17] Passow, Nos. 511, 512. A similar story to this one is attached to the castle of Scodra. *See* Mackenzie and Irby's 'Travels in Turkey,' pp. 535, 544.

lage finds his wife washing clothes at the fountain, and, after testing her fidelity, is put through a searching cross-examination by that lady, who seems accustomed to receive all such approaches with disfavour. He enumerates a variety of objects by which his home may be distinguished, but in no case succeeds in satisfying his questioner, until he has named some body-marks which are known to him alone, on which he receives the acknowledgment he deserves.[18] Another, called "The Abduction," which Fauriel regards as one of the most ancient and at the same time of great poetical merit, describes the stratagem by which a lover recovers his lady, who is being married to a Turk against her will in his absence.[19] Though at a great distance from her house, he discovers, by the neighing of his horse and other omens, what is about to happen. He rides to the spot with lightning speed, and on his arrival at the door it is again the neighing of the steed that calls the attention of the lady, who is on the point of being given away. In order to deceive her new suitor, she explains that it is her eldest brother, who has brought her dowry, and goes out to meet him and offer him the stirrup-cup. As soon as she approaches his side, he catches her up, nothing loth, places her before him on the saddle, and is out of sight long before the Turks have seized their guns to give chase. Besides an element of the marvellous, which runs through this and many other poems, the probabilities of the case are here more than usually ignored, for the hero is represented, notwithstanding his haste and the urgency of the circumstances, as stopping on two separate occasions, when on his way to effect the rescue, in order to visit his father and mother in different places, and to obtain from them

---

[18] Passow, Nos. 441 foll.   [19] Ibid., No. 439.

respectively an augury as to the success of his enterprise. According to our notions, the introduction of such things would mar the completeness of the poem, but instances of filial affection are a subject on which the Greeks are always glad to dwell, and they feel no objection to sudden transitions or violations of poetical unity. Together with these ballads we may mention, as turning on real or supposed events, the stories of death by drowning connected with the river Peneius;[20] that of the Charamides, or brigands, where one brother is killed by another, and recognised by him at the moment of death;[21] and those relating to the "Beauty of the Tower," which have been noticed in connection with the Vale of Tempe.[22]

The dirges or myriologues ($\mu\nu\rho\iota\circ\lambda\acute{o}\gamma\iota\alpha$, from $\mu\acute{\nu}\rho\circ\mu\alpha\iota$), of which we must now speak, are a class of compositions intended for recital at funerals. These are always composed and sung by women, and according to Fauriel they are almost universally improvised. About this latter point, however, there is considerable doubt, and though from time to time it certainly happens that they are the effusions of the moment, a thing which is facilitated by the habitual use of the regular ballad metre, and by the introduction of "common-place," yet it seems now to be thought that in the majority of instances the pieces are known beforehand. In some places there are found women who are professed myriologists, or at all events are famed for their skill in that art, and are commonly invited to take their share in funeral ceremonies: and these persons may be heard, when at their work in the fields, practising dirges on ideal subjects. On the occasion of a funeral the female relations of the deceased, together with any others who may happen to be present,

---

[20] See above, p. 71.   [21] Passow, No. 487.   [22] See above, p. 68.

one after another, according as their feelings prompt them, utter their laments over the coffin, which is always left open in Greek burials, so that the face of the dead is exposed to view. These songs are recited slowly to a peculiarly plaintive melody, and continue during the funeral procession, and until the body is lowered into the grave, only ceasing while the prayers are being chanted. In cases where the effusion is extemporized, the strain on the mental faculties, combined with the excitement of the feelings, has so overwhelming an effect on the reciter as to cause her in some instances to faint away. The following remarkable account of such an occurrence has been given by M. Fauriel :—

"A woman of Metzovo, on the Pindus range, aged twenty-five years, had lost her husband, who left her with two children of tender age. She was a poor peasant of very simple character, who had never given marked proofs of talent. Leading her two children by the hand, she came into the presence of her husband's corpse, and commenced her myriologue by reciting to the dead a dream which she had had several days before. 'I saw the other day,' she said to him, 'at the door of our house a young man of lofty stature and threatening aspect, with white wings expanded on his shoulders; he was standing on the threshhold of the door, with a drawn sword in his hand. "Woman," he inquired, "is thy husband within?" "He is," I replied; "he is combing the hair of our little Nicola, and caressing him to prevent him crying. But come not in, thou terrible youth; come not in; thou wilt frighten our child!" Yet the young man with white wings persisted in his desire to enter. I would have forced him to retire, but had not the strength. He rushed into the house; he threw himself on thee, O my beloved, and smote thee with his sword. He smote thee, unhappy one! and behold, behold thy child, our little Nicola, whom he would have killed also——'

"After this commencement, the tones of which, no less than the words, had caused the bystanders to shudder, some of whom looked towards the door, as if to see whether the young man with white wings were still there, the others at the little child clasping his mother's knees, she threw herself sobbing on the corpse of her husband. It was with difficulty that they tore her from it, and no sooner was she removed than in a fresh transport of emotion she recommenced her address. She asked her husband how she was henceforth to support her children and live herself? She recalled their years of married life; the fond love she had borne to her spouse, the tender care with which she had brought up her children, and ceased at last only from exhaustion, in a fainting state, and resembling in her pallor him to whom she had been addressing those mournful words." [23]

Of the myriologues which have been collected very few can be regarded as extempore effusions. Most of them bear evident traces of careful composition, and, from the nature of the case, poems composed under the influence of great excitement are too evanescent to be remembered afterwards. Some of those which are in rhyme are highly poetical; but for dramatic treatment and imaginative power, hardly any is equal to the ballad of "The Spectre," which has been given above to illustrate the belief in the Vrykolaka, if, as is not improbable, that piece is to be classed as a myriologue.

Closely connected with these, again, are the numerous poems relating to Charon or Death, that fearful being (for it is much too real to be called a conception) who has taken so permanent a hold on the imagination of the Modern Greeks. I should be anticipating what I have to say on this subject under the head of the popular mythology, if I were to describe the beliefs that prevail concerning him: it may suffice for the present to introduce one or two of the pieces which relate to this subject. In some of these he is represented as wreaking his vengeance on a young and beautiful girl, who in a proud moment has dared to speak slightingly of his power; in others on shepherds or lusty warriors who dispute with him the possession of their lives. Many also are the tales of the under-world, and of adventurous attempts to escape

---

[23] Fauriel, 'Chants Populaires,' i. p. cxxxvii.

from that dreary prison. But the object of all is to exalt his irresistible might, which neither beauty nor strength nor daring can escape; and his unsparing, merciless character, which was never diverted from its fell purpose. The following poem, in which his mother, who represents female tenderness and compassion without any further allegorical significance, is described as interceding with him, is intended to express especially the latter element in the popular belief:—

"CHARON AND HIS MOTHER.[24]

" Out in the moonlight Charon stood; his courser he was shoeing:
When thus his mother spake to him, and thus a boon she asked him:
'My son, thou'rt going to the chase, thou'rt going to thy hunting;
Take not the mother with her babes, nor brothers with their sisters.
Take not the newly-married bride, who wears her wedding garland.'
'Where I find three, two are my prey; where two, one is my portion;
And where I find but one alone, yea upon him I fasten!'"

The next, which called forth the admiration of Goethe, and is worthy of being illustrated by the pencil of Gustave Doré, represents him on horseback, carrying off the souls of the dead.

"CHARON AND THE SOULS.[25]

" Why are the mountains overcast? why are the hill-tops shrouded?
Is it the wind that vexes them? is it the rain that smites them?
'Tis not the wind that vexes them, 'tis not the rain that smites them;
'Tis Charon, who is sweeping by with souls of the departed:
The young he drives before his face, the old he drags behind him,
The children to his saddlebow, the tender ones, are fastened.
On bended knees the young implore, the aged men beseech him:
' Good Charon, at a village halt, halt at a cooling fountain;
That th' old may drink a quenching draught, the young with quoits amuse them,

---

[24] Passow, No. 408.   [25] Ibid., No. 409.

And the children, too, the little ones, may gather knots of flowers!'
'At ne'er a village Charon halts, at ne'er a cooling fountain;
Mothers would come for water there, and recognise their children,
And wives their husbands meet again, and never more be parted!'"

From these we pass to the nuptial songs and love poems, which are the most finished in point of composition, and the most varied in treatment, of any in the collection. This is to be accounted for by their being mostly drawn from the cities, and from the coasts and islands, where higher civilization is found, though a few are derived from the country districts. A large number are rhymed, and the metre assumes a variety of forms, instead of being confined to the traditional type of the Greek ballad verse. This remark is especially applicable to the wedding songs, which are sung in chorus, and often antiphonally by the men and women, as was the case with some of the hymeneal songs of classical times; not unfrequently, too, one set of voices catches up the other in the middle of a sentence, and the meaning is bandied to and fro in a somewhat intricate manner. Each of the numerous domestic ceremonies which precede and follow the marriage rite has a song appropriated to it, while others are used as accompaniments to the dances which succeed. The following little piece, in the ordinary ballad metre, is used in Thessaly, according to Fauriel, for the latter purpose. There is a passage in it which may remind the reader of something in an English poet, whom Manzoni has called "un barbaro chi non era privo d' ingegno."

"Come hither, youths, and join the dance, and sing the song in measure;
Come sing the song that sweetly tells how fancy is engendered:
It is engendered in the eyes; upon the lips it blossoms;
Then downward to the heart it glides, and in the heart is rooted."[26]

---

[26] Passow, No. 528. Compare Fauriel, ii. p. 234.

The next, a wild and passionate love poem, which is likewise used for dancing to in Thessaly, was probably composed in the mountains, as the subject of it, Demos, would seem to have been the captain of a band of Armatoles. This name was well known on Olympus; in fact, the person here mentioned may have been the identical Captain Demos whose skill in carving a lamb and bleeding a woman with the same weapon Mr. Urquhart has so humorously described in the course of his rambles on that mountain in the 'Spirit of the East.'[27] Like the preceding one, it is not rhymed; but the metre in which it is composed is very graceful, and is not found elsewhere: it has the peculiarity of requiring that the name of Demos should be introduced in the first line of each couplet.

DEMOS.[28]

"Thine eyes, O Demos, with their lovely light,
    Thy finely-pencilled eyebrows,
They make me swoon, O Demos, as I gaze,
    They make me die with longing.
Draw thy bright sword, O Demos, from its sheath,
    And plunge it in my bosom,
And catch the blood, O Demos, as it flows,
    In a gold-broidered napkin;
Then bear it, Demos, to thy cantons nine,
    To thy ten village-districts,
And when they ask, 'What's this?' O Demos, say,
    'The blood of my beloved.'"

We may now take a specimen of the more finished love poems, which are in rhyme, and in more elaborate metre. In these the influence of Italian poetry is very directly traceable; and in the islands it is not surprising to find that this is the case, considering the amount of communication that has long existed between the two

---

[27] Vol. i. p. 451.   [28] Fauriel, ii. 154.

countries, and the widely-extended use of the Italian language in the Ægean.[29] For this class of compositions the modern Greek tongue, when somewhat refined, is admirably adapted, owing to its sweetness, flexibility, and liquid clearness, together with an element of unaffected and truthful simplicity, which arises from its inartificial character. And though in some of these points it is inferior to the Italian itself, yet it has one great advantage over it and most other European languages in the frequency of the oxytone accent, which imparts to it, as it must also have imparted to ancient Greek, a peculiar brilliancy and expressiveness. The poem which I have chosen, entitled "The Garden," is from Smyrna; and the occurrence in it of occasional classical words, which would not be found in the Romaic of the country districts, would be sufficient to show that it was derived from a place of some literary culture.

> " Within a fair plantation
> Enamelled o'er with flowers
> I passed the early hours
>   My anguish to remove;
> To chase the thoughts that vexed me,
> And fancies that perplexed me,
> For all the pain I suffer
>   From the sweet maid I love.

---

[29] The latest addition that has been made to this class of poems is of an interesting character, as it consists of Greek songs which have been discovered in South Italy, where a number of Greek colonies existed during the Middle Ages, the inhabitants of which spoke Greek till lately, though now they speak Italian. In the 'Saggi dei Dialetti Greci dell' Italia Meridionale,' edited by Domenico Comparetti (Pisa, 1866), upwards of forty of these songs are given, which are almost all in single stanzas of eight lines, in rhyme. Here and there they are intermixed with Italian words, and at first they had to be written down in Roman letters, and in some cases from a corrupt pronunciation, and then interpreted into their Greek form. All things considered, they are more accurately preserved than might have been expected.

"And lo! as there I wandered
 Within that fair plantation,
 And paused in admiration
  Of the blossoms, fair to see,
 A bird that shyly cowered
 By citron leaves embowered,
 Was singing, as I listened,
  A pleasing melody.

"For ever, while it warbled,
 I seemed to hear it saying,
 'Young men, avoid delaying,
  Full soon your joys are o'er:
 And you, fair maids, go marry;
 Be wise, nor longer tarry,
 For time is ever flying,
  And will return no more.'"[30]

A word must be added about the distichs, upwards of eleven hundred of which are appended to Passow's collection. These, with few exceptions, are devoted to amatory subjects, and are of very various poetical merit, the valuable ones being rather the exception. They are universally rhymed, and almost all are in the ordinary ballad metre. A few of the best have been selected for translation, the last being a double distich.[31] Let us begin with one which might serve for a description of the Park in June.

DISTICHS.

I.

"'Tis now the joyous summer, when all nature's a delight,
 And golden girls are seen abroad in dainty dresses bright."

II.

"Though lovely all thy features are,
  Fairest to me thine eye,
 Where moonlight and the morning star
  Meet in a mellow sky."

---

[30] Passow, No. 586.
[31] Ibid., Nos. 1069, 95, 330, 389, 379, 705, 772.

III.

"Thou art the key of heav'n above, on earth beneath a fountain,
Among the maidens fair and tall, as cypress on the mountain."

IV.

"O that, like one who reads the skies,
    By gazing I might find,
Whose image dwells within thine eyes,
    Imprinted on thy mind."

V.

"O bitter is the salt sea-wave, but sweet that flood will be,
Ere the wound heals, the cruel wound, which I received from thee."

VI.

"He that with blood from forth his heart
    An image fair doth trace,
Painteth with skill surpassing art
    What he shall ne'er efface."

VII.

"Lovely bird with painted plumage, that dost haunt the tangled grove,
I am setting nets and springes to entrap thee with my love:
And, if in my nets I catch thee, with a faithful heart I swear,
In a golden cage I'll keep thee, lovely bird with plumage rare."

The metre in which the ordinary Greek ballads are written is that which is called the *versus politicus* (στίχος πολιτικός), *i. e., popular* verse, a long and balanced measure, regulated entirely by accent, without any regard to the quantity of the syllables. In the hymns of the Greek Church, from which source is derived whatever slight evidence exists as to the change that passed over the national poetry in the early part of the Middle Ages, specimens of accentual verse of one sort or another may be found as early as the eighth century;[32] but the first

---

[32] *See* the Introduction to Sophocles' 'Dictionary of Later and Byzantine Greek,' p. 65.

examples that we find of that particular form which was called *political* are of the middle of the eleventh century, though the name itself implies their previous existence, as being commonly found at that time in the mouth of the people.[33] In the following century Tzetzes and other writers are known to have composed poems of considerable length in this metre, though Tzetzes apologises for its use as a vulgarism by saying, that in that age there was no distinction between what was classical and what was barbarian. After the time of the Fourth Crusade and the occupation of Constantinople by the Latins, when the minds of the Easterns were impregnated with ideas derived from the spirit of chivalry, this same verse was employed in the popular romances which were composed in imitation of the literature of the West.[34] From that time to the present day its use seems to have been perpetuated both in written and unwritten compositions. To describe somewhat more exactly the metre itself—it is the iambic tetrameter catalectic, with a break after the first dimeter, which divides the verse into two hemistichs; and iambic feet are allowed to be replaced by trochaic with about the same amount of licence as is admitted in ordinary English heroic verse. In the early form of the verse, as it appears in Byzantine writers, this license seems to have been confined to the first foot of each hemistich, but in the modern verse it has extended further. It has sometimes been called a fourteen-syllabled iambic foot with a double ending;[35] but though

---

[33] See Thiersch, 'Ueber die Neugriechische Poesie,' p. 13, note 9, where he shows that the word πολιτικός in ancient Greek, and *publicus* in Latin, were used in the sense of "popular" or "vulgar."

[34] On the influence which French literature exercised on the Greek writings of that period an elaborate essay has lately been published by M. Gidel, entitled 'Etudes sur la Littérature Grecque Moderne.'

[35] Professor Blackie's 'Homer,' i. p. 420.

this is true in itself, yet in the modern Romaic verse the amount of elision and coalescence of syllables that is allowed in pronunciation is so large, that any description of the metre according to syllables is better avoided. In English it is not, in this exact form, the commonest of the ballad metres, yet it is found from time to time, as in the following instance from the ballad of "Sir Lancelot du Lake":—

> "Then into England straight he came
> With fifty good and able
> Knights, that resorted unto him,
> And were of his round table."[36]

Anyhow it is closely allied to the corresponding metres which are found to prevail in the ballads of all nations, from the Saturnian of the Romans downwards, so that it appears to be the most natural of all metres.[37]

The use of rhyme, which we have seen to be common in the more finished poems, is thought to have been introduced from Italy at the time when the Morea, Negropont, and other islands, were subject to the Venetians. Occasionally, indeed, it is found in early hymns of the Greek Church,[38] which is not surprising when we consider how early it was introduced into the corresponding compositions in Latin: but it appears to have been discontinued, and is not found in the mediæval Greek writings.

---

[36] Percy's 'Reliques,' i. p. 216. The early English poem called the "Ormulum," is entirely composed in this metre. Dr. Guest suggests, though with what likelihood I know not, that the political verse was borrowed from the Greeks by the English in the twelfth century, and that it had great influence in forming other English metres.—'History of English Rhythms,' ii. p. 184.

[37] For this reason it might be more accurate to regard it as a trochaic measure, that being the foot which is usually found in ballad poetry; but it is simpler to speak of it as iambic.

[38] See, for instance, Sophocles' 'Dictionary,' p. 60.

The earliest poem which is found in regular rhyme is the 'Description of the Plague of Rhodes,' published in the year 1498, by Emmanuel Georgilas, a native of that place; this, however, cannot be regarded as a popular poem. That the ordinary use of rhyme was of Italian origin is further confirmed by the words used to express it in modern Greek (ῥημάδα, ῥημαρίζω), which are evidently derived from *rima*, *rimare*, in Italian. On the other hand, the frequency of its occurrence, especially, as we have seen, in the distichs, which are so numerous and drawn from such a wide area of country, sufficiently proves that it is not, like the Italian metres, of late introduction, though a peculiar application of its use in connection with those metres may be so. Its employment is facilitated by the system of inflections remaining to a great extent in the verbs, which causes the same forms and the same sounds to recur, just as in Homer the *homœoteleuton* is found to arise from this cause, probably without any intention on the part of the poet, and perhaps unavoidably. The plastic condition of the language also, which is not as yet stiffened by conventional rules and the influence of a written literature, and consequently admits of change and adaptation in the terminations, renders its introduction more feasible, in the same way as the similar condition of Italian in Dante's time rendered more manageable the intricacies of the *terza rima*. To these influences may be added the license which is admitted in shifting the accent for purposes of scansion and rhyme, especially when it is thrown forward on to the last syllable.

The manner in which these ballads are sung is very peculiar, and very distasteful to a Western ear. The airs to which they are set are monotonous, and as a rule very high pitched, and the words are drawled out slowly, and the syllables repeated in an extremely wearisome manner, in addition to which words and exclamations are frequently inserted as a sort of refrain. As the Greek church music, which is certainly of great antiquity, closely resembles this as well in sound as in the mode of enunciation, and both of them are more like the ecclesiastical plain chant than any other music in Western Europe, it probably dates from very early times.[39] Of its origin, however, nothing is known, though the study of it might throw light on many difficult points connected with ancient and other music. Shocking as it may be to all our preconceived ideas, yet if we knew more about the music of Greece in classical times, and could speak with more confidence of the intervals and combinations of sounds which have been admitted in different ages and countries, we might perhaps discover that the melodies of the most refined nation the world has ever seen were nearly akin to these raw, and to our ears inharmonious strains.

There is an additional point of view, and one of great interest, from which these ballads may be regarded, namely, as furnishing us with a means of forming an estimate of the character of the people amongst whom they have originated. Prescott has somewhere remarked, that for the production of a fine ballad-poetry in any country two conditions are requisite: first, that the nation should possess depth of character; secondly, that it should grow up in the midst of stirring circumstances:

---

[39] Karl the Great is said to have been profoundly affected by the solemn music of the Greek service. Yet after the capture of Salonica by the Normans, in 1185, among the insults which were heaped upon the inhabitants, it is mentioned that when the Greek priests chanted their service, the Norman soldiers howled out a chorus in imitation of beaten hounds. This is a sufficient proof of the nasal character of the ecclesiastical music of that period. See Finlay's 'Byzantine Empire,' i. p. 150; ii. p. 265.

and in this way he accounts for the fact that the ballads of Spain and England form the finest collections that modern Europe has produced. Conversely we may argue that wherever a remarkable ballad literature exists, or legends and traditionary stories capable of forming the groundwork of such a literature, the character of the people must be elevated and marked with salient traits. Thus among the early Romans the stories of Mucius thrusting his hand into the flames, and of Clœlia swimming the Tiber, though we may no longer be able to regard them as historic facts, are a sufficient evidence that those who originated and cherished them possessed more than ordinary qualities, and a capacity for future greatness. With regard to the modern Greeks, the same thing appears from an examination of this collection; and as we read some of the finer pieces that have been composed in the country districts, we cannot help feeling the greatness of an illiterate and uneducated people, who can either give birth to, or listen with enthusiasm to, such compositions. It may be worth while to mention some of the more noteworthy points in which this is traceable.

On the independent spirit of the people, their love of freedom, their heroism and intrepidity, their intensity of national feeling, there is no need to dwell after what has been said of the Cleftic ballads, which are nothing else than a history of the protracted struggle for liberty against the Turks. Nor again is any proof required of their imaginative power, in addition to those furnished by the more ideal poems which have been noticed. It is rather those qualities that lie beneath the surface to which our attention may be most profitably given. Such, for instance, is the strong family feeling which is in such a high degree characteristic of the Greek people. Some of the observances which are closely connected with their domestic life—such as the ceremonies connected with marriage and death, and the expressive songs which accompany them—may, no doubt, be paralleled by what is found among other nations; but there is one custom with which many of the ballads are associated which throws especial light on this feature of the Greek character. These are the observances connected with the departure of any member of a family, whether for purposes of education or trade, or to avoid persecution, to those hated foreign countries to which the epithet of "desolate" or "lonely" (ἔρημα) is constantly applied. On such occasions the person who is about to leave assembles his relations at a feast, where songs appropriate to the occasion are sung, and, when the time for his departure is arrived, he is escorted by the whole company for several miles from his home, until at last, at the moment of separation, he recites to them a piece—either one of those already existing on the subject, or one composed expressly for the occasion—in which the bitterness of leave-taking and the miseries of a foreign home are passionately expressed. Many of these are given under the head of *Carmina Hospitalia* in Passow's collection. Amongst them is one, which is repeated in a variety of forms, supposed to be sung on such an occasion by a son, who has been slighted and illtreated at home by his mother, in which he represents her as driving him away, and describes her anguish when she hears that he has been killed in a foreign land, and become the food of birds of prey. It is related by Fauriel that in one instance, the facts of which he had investigated, a mother who was addressed in this way was overcome with emotion in the course of the recitation, and at its conclusion, throwing herself on her son's neck, entreated his pardon for her former conduct, promising at the same time to deal with him as a loving mother for the future.

As might be supposed, where the domestic feeling is so strong, the conjugal tie is regarded as pre-eminently sacred. This is evidenced by such poems as those which relate to the recognition of husbands by faithful wives after long absence; and the other poem referred to above, called "The Abduction," implies something of a chivalrous relation between the sexes, which is fully borne out by the love poems. In these are depicted all the varied emotions, all the lights and shades of passionate feeling, which belong to a quick, tender, and sensitive nature. One favourite theme is the story of the husband who, on returning to his young wife after a short absence, finds that she has died in the interval. She is represented as giving a dying charge to her mother to console and comfort him; but when on his arrival he finds the priests preparing for the funeral, and the grave being dug, and discovers what has happened, he requests the gravediggers to make it large enough to receive him also, and immediately draws his dagger and plunges it into his heart. Then it is described how, after their burial, the plants which spring from their tombs wave in the wind, and when Sir Boreas (ὁ κὺρ βοριάς) blows loudly they mingle their branches, so that they who were not permitted to be one in life, are united after death.

Again, to turn to another phase of pathetic feeling, what can be more touching than the ballad of "The Young Sailor," which is frequently sung by mariners in the Ægean? It begins by a short prologue describing the miseries of a life on the sea, and then passes to the subject of the story, who is represented as feeling himself dying as he enters the haven, within which lies his home. When the captain desires him to get up and take the bearings for them, the young sailor begs them to lift him into a sitting posture and bring him the compass, and then, pointing to a mountain whose summit is seamed with gorges (an excellent description, by the way, of the limestone peaks of Greece), tells them to make for it and cast anchor at its foot. Then, having performed his duty, he asks as a last request that when they reach the land they will bury him on the shore, that he may still hear the voices of his comrades, and their familiar cries which he loves so well:

> "A boon of thee, my mate, I crave; a boon of thee, my captain:
> Lay not my bones in holy ground, in church or monastery;
> But on the margin of the beach, beneath a bed of shingle,
> That there the sailor-lads may come, and I may hear their voices." [10]

It is somewhat surprising at first sight to find a people who have so strong a sense of pathos, almost entirely destitute of humour, which is so closely allied with it. Perhaps, however, the truth is that neither the nation itself nor its literature has reached the stage of observation and criticism which humour implies. No one who has associated with the Greeks would suspect them of being wanting in fun, and this breaks out here and there in the ballads, especially in the form of amusing banter against the monks. Again, their quickness and vivacity of temperament cannot fail to give birth to witty sallies, as is the case among the Irish. But of real humour, that is, the power of seeing the incongruities of things, there is hardly any trace in their compositions. One cause that may have tended to produce this effect is the sad and serious condition of a people conscious of living under

---

[10] Passow, No. 391.

oppression and in a state of insecurity, which, notwithstanding natural liveliness, makes itself at once felt whenever reflection is involved, as is necessarily the case in poetry.

To take one more point in their character which these poems illustrate: there is constant evidence of their possessing what an imaginative temperament, when placed in the midst of striking features of natural scenery, almost necessarily involves,—a strong sympathy with external nature.[41] Thus in the Cleftic ballads, when the death of one of their leaders or other bad news is to be announced, the sky is overcast, the rain falls, and the torrents descend, in correspondence with their feelings; sometimes, too, it is the sighing of the wind in the trees, or its roaring round their mountain abodes, which is the presage of ill fortune. The same thing is implied in those poems where allegory is introduced; while in others there is a connection implied (though this idea may perhaps be of mythical origin) between the life of a man and that of a plant or tree, the duration of one depending on that of the other.[42] In a love-song brought from the isle of Melos a fruit-tree is described as withering in an orchard without any apparent cause, and when its companions inquire what ails it, it replies that two lovers had been married beneath its shade, and there had sworn never to be separated, but now their vow was broken, and therefore its branches wither.[43] And in a variety of ways the "pathetic fallacy" is introduced, and human feelings are attributed to inanimate objects and

---

[41] The ancient Greeks are not in reality an exception to this rule, for it was not until they were engrossed by city life and its associations, that they seem to have lost the conscious admiration of scenery.

[42] *E.g.*, Passow, No. 153, one of the prettiest Cleftic ballads.

[43] Passow, No. 661.

to animals. The latter of these are frequently made to play a very prominent part in what is represented as happening: the birds are the interpreters of fate, being either introduced to bring the news of some occurrence, or acting as a sort of chorus, which comments on what is passing, and indicates the point of view from which the audience are intended to regard it. The dying Cleft expresses a wish that an opening may be left in the side of his tomb, by which the birds may fly in and out, and bring him news of the outer world. Nor is this sympathy confined to them. When Demos is lying outstretched on the plains of the Vardar, it is his horse that urges him to rise and follow the rest of his company; and when he feels his strength is failing, he commits to him as to a faithful friend the ring and other tokens which are to be borne to his lady-love, and bids him to dig for him a grave on the spot with his silver-plated hoofs. In these and innumerable other instances the marvellous element is introduced with such perfect simplicity, and withdraws the narrative so completely from the course of ordinary occurrences, as to appear perfectly natural, and by no means to outstep the license of poetic treatment.

And now, to conclude this imperfect sketch of a very interesting subject, I would ask my readers whether they remember in their younger days ever to have heard anything of which the following Greek nursery rhyme may remind them. If so, they may perhaps find it a pleasant subject of speculation, whether such resemblance is to be attributed to the tendency of the human mind to give birth to similar products under corresponding circumstances in widely different localities, or whether it is to be explained by the supposition of the existence of some primæval nursery-rhyme of this form, which was sung during the infancy of nations.

"There was an old Man, and he kept a cock, which crowed and waked this lonely old man.

There came by a Fox, and ate the cock, which crowed and waked the lonely old man.

There came by a Dog, and ate the fox, that ate the cock, that crowed and waked the lonely old man.

There fell down a Log, and killed the dog, that ate the fox, that ate the cock, &c.

There came a Furnace, and burnt the log, that killed the dog, that ate the fox, &c.

There came by a River, and quenched the furnace, that burnt the log, that killed the dog, &c.

There came by an Ox, and drank up the river, that quenched the furnace, &c.

There came by a Wolf, and ate the ox, that drank up the river, &c.

There came by a Shepherd, and killed the wolf, that ate the ox, &c.

And last came the Plague, which carried off the shepherd, who killed the wolf, that ate the ox, that drank up the river, that quenched the furnace, that burnt the log, that killed the dog, that ate the fox, that ate the cock, that crowed and waked the lonely old man."[41]

---

[41] Passow, No. 274.

CHAPTER XXIX.

THE MODERN GREEK POPULAR TALES.

Difficulty of discovering the Greek Stories — Mode of collecting them: General Remarks on Popular Tales — Their Characteristics — Beast Fables — Resemblances between the Tales of various Countries — Cinderella — "The Snake-Child" — Norse and Greek Tales — Community of Origin — How explained — European Tales not derived from India — Greek Tales not borrowed — Indo-European Origin — Mythical Character — Interpretation of "The Sleeping Beauty" — Various Forms of the Story — The Ideas embodied in the Tales few and simple — Various Modes of their Formation — Peculiarities of the Greek Stories — The Drakos — Similar Conceptions — "Lazarus and the Dragons" — The Drakos Mythological — Represents the Thunderstorm — Ballad of "Jack and the Drakos."

ONE characteristic of the Greek ballad poetry, of which we have just been speaking, is its publicity, and the absence of any attempt on the part of the people to conceal it from strangers. Though the songs have never been committed to writing, yet they are recited on festive occasions, and in the presence of great multitudes, where no one is excluded; besides which, they may constantly be heard, as they are chaunted by the people, on the mountains, by the roadsides, and in the towns and villages. With the literature of which we are now about to speak the case is exactly the reverse; and while it is in the power of any one who will to write down the ballads from dictation, the Popular Tales are kept so carefully concealed, that even a careful investigator might pass through the country without discovering a trace of their existence. The collector to whose industry we are indebted for the large number of these stories which have lately been published,[1] assures us that, though in

[1] Von Hahn, in his 'Griechische und Albanesische Märchen.'

the course of twenty-seven years' residence in the Levant he had been in constant intercourse with the Greek peasants, had eaten and drank, hunted and travelled with them, and slept for weeks together in their cottages; though he had passed many an evening in Greek, Albanian, and Bulgarian khans in the company of other travellers round the common fire, and many a day in small coasting vessels crowded with passengers; yet, notwithstanding various attempts, he had never succeeded in eliciting from them a household story. His experience in this respect coincides with that of all other persons who have endeavoured in different countries of Europe to disinter this strange literature, which lies buried in the memories of the people. The utmost caution and the greatest intimacy are always requisite, in order to overcome that innate shyness which prevents them from communicating it to others not of their own class. Few phenomena, indeed, are more extraordinary than the existence among the lowest ranks of all nations of a class of compositions of great beauty and imaginative power, which, nevertheless, was unknown until within a few years, and seldom fails to vanish before the approach of education.

The treasures of this class which exist in modern Greece were brought to light by Von Hahn in the following manner. Being aware, from the experience of other investigators, that women are the chief depositories of these tales, he looked about him for some intermediate agency by means of which to extract them from this source. Accordingly, when he was residing as consul at Yanina, having become acquainted with the pupils of a Greek school in that city, he persuaded them, by the offer of a small remuneration, when they returned for the holidays to their respective homes, to get their mothers and sisters to relate to them the stories that they knew, and to write them down from their dictation, cautioning them at the same time to avoid all amendments or alterations of the stories from their original form. Again, when he was transferred to the consulate of Syra in the Archipelago, he employed a Greek woman who could write to collect for him; and in the same place he obtained other specimens by the help of Greek ladies of the upper classes. To these was added a collection made by a Swiss professor in the north of Euboea. In this way he gradually brought together about one hundred Greek and twelve Albanian stories, which he subsequently translated into German. They prove to correspond in the most remarkable manner with similar collections which have been made in other countries, and in many respects throw considerable light on the whole subject of Popular Tales. But before proceeding further, it is necessary to say a few words on the question generally, as what follows would otherwise be unintelligible to those of my readers who have not turned their attention to this branch of literature.

From the time that the brothers Jacob and William Grimm began to collect the stories that exist among the peasants in Germany, and their example was followed by numerous collectors in other parts of Europe, it has been known that everywhere the lower classes are acquainted with, and able to recite, tales of a very peculiar character, and that these are universally marked by the same characteristics. If, before examining the contents of the collections, we were to hazard a conjecture as to the nature of these stories, we might perhaps fancy that they would have reference, more or less, to the circumstances of the times and the people, or would contain narratives of past events perverted and disguised. As a matter of fact, we find nothing of the kind: they are purely ideal and imaginative. The times to which they refer never existed at all; and the world which they describe is peopled with kings and queens, princes and princesses, with adventurous heroes and ill-fated heroines, with giants and dwarfs, genii and fairies, with dragons, serpents, and talking animals. Among the characters whose histories are related the most common are clever Tom Thumbs, half-witted simpletons, bold rhodomontaders, the clever thief, the practical joker and humorous cheat. The points and incidents on which the stories turn are transformations and metamorphoses of various kinds; staircases leading to magic regions below the earth; caps, cloaks, and rings, which render the wearer invisible; the power of obtaining anything by wishing; enchanted horses, which are summoned by a given signal to the succour of their owner; maidens exposed in wildernesses or on the sea, frequently by the malice of a wicked stepmother, and saved by some supernatural agency; the value attached to the number 3, which makes the youngest of three brothers or sisters the best and most fortunate, and the third time always lucky; together with a number of other features usually found in fairy tales and children's stories, all of which have originally come from this source. But in all of them the marvellous element is an essential ingredient: where it is not found, the story is no longer, in the technical sense of the term, a Popular Tale. Together with this, all kinds of inconsistencies are admitted, though in some cases these are modified and softened down, in order to avoid the appearance of absurdity; in this way the existence of many such incongruities is an argument in favour of the more primitive character of the story in which they occur.

As the Popular Tale is thus marked off by features of

its own from ordinary stories, so on the other side it is distinguished from that, to which it is also closely related, the Myth, or Saga. Both of them deal with the marvellous and supernatural, but the myth, as it is exhibited in the mythology of Greece and of the north of Europe, attributes to the gods the influences that are at work in the world, while in the popular tale the gods are altogether excluded, and the power is referred to the lower agency of fairies and demons. Again, the myth is more nearly allied to the domain of history, the popular tale to that of poetry. The former is more or less attached to facts and places; and whatever wild forms the superstructure may take, it is based on what are conceived of as existing realities: its Apollo and Thor are regarded as actual divinities; its Achilles and Arthur as heroes who once lived and breathed. The latter is absolutely unattached, it is fixed to no particular locality, its personages do not necessarily possess any distinctive names; the world in which it exists is beside and altogether apart from the world of actual occurrences. Thus, for instance, when a story commences in the following manner—"When Laomedon built the city of Troy, Poseidon and Apollo were doomed to serve him for wages," we know that we are to have a myth; but when the beginning is such as—"There was once a king and queen, who had twelve children," we may expect a popular tale.

One particular form of these tales requires to be noticed separately, namely, the Beast fable. This attains its most perfect development in the great Beast epic of the Middle Ages, the story of "Reynard the Fox": but in a more fragmentary shape it is found in some other countries, and in particular among the modern Greeks. In it the animals are treated of quite independently of their relations to men; but, while the peculiar character of each species is maintained,—such, for instance, as the cunning of the fox, and the clumsy ferocity of the wolf, and even minor traits, like the bear's fondness for honey,— yet they are made to act and speak in all respects as if they were human beings; nor is there felt to be any incongruity in the murdered hen, in the story of Reynard, being brought into the presence of the king on a bier, or in the wolf receiving the tonsure when he betakes himself to a monastery. In this way the Beast fables are a sort of reflection of phases of human life, a description of the life and habits of animals by means of ideas drawn from the actions of men and the customs of society, but without any satirical or didactic purpose— in which respect they are to be distinguished from the Æsopic and similar fables of later date, where the object is to draw a moral. The original stories took their rise in the intimate relations which existed at a very early period between men and animals, and the close observation of their habits which was only possible under such circumstances. The interest which arose from continually watching them caused the primitive races to attribute to them a community and a history like their own, and these ideas embodied themselves in stories like those that have come down to us. The following, which may be taken as a fair specimen of those that occur in the modern Greek collection, closely corresponds to one of Grimm's German tales; where, however, the cat and mouse take the place of the fox and wolf, and the names of the children are given as "Skin-off," "Half-out," and "Quite-out."[2] In a Celtic dress the same story appears as "The Wolf, the Fox, and the Keg of Butter," and the

[2] Grimm, 'Märchen,' No. 2.

names are "Under its mouth," "Half-and-half," and "Licking all up."

"THE WOLF, THE SHE-FOX, AND THE POT OF HONEY."

"There was once a wolf, called Mr. Nicholas, and a she-fox called Mrs. Maria. These two bought a field between them, and prepared to till it. In the way of provisions, they took with them a pot full of honey and a basket of white bread, and when they had concealed their victuals they set to work to dig. After digging a good while, the fox began to feel hungry; but, as she was ashamed to tell Mr. Nicholas, she bethought her of a device, and pretended she was being invited to a christening, exclaiming, 'All right! all right! I'm coming!'

"On hearing that, Mr. Nicholas asked, 'Who's calling you then?' and she replied, 'They're calling me to stand godmother to a child; so I'll go there, but I'll be back in a minute.'

"'Go along then,' said Mr. Nicholas; 'but take care you come back soon.'

"So Mrs. Maria started off, stole to the place where the pot of honey and the loaves of white bread were concealed, and ate till she could eat no more.

"Then she returned to Mr. Nicholas, and he said to her, 'Welcome, Madam Gossip; and what name have you given the child?'

"'Firstling,' was her answer.

"So they betook them again to their work; but after a while Mrs. Maria began again to hanker after the honey, so she cried, 'All right! all right! I'm coming!'

"Then Mr. Nicholas asked, 'Well; who's that calling you now?'

"'Why, I have to stand godmother again.'

"'Go along, then; but take care you come back soon.'

"So Mrs. Maria stole off to the honey, ate till she could eat no more, and then came back to the field.

"Mr. Nicholas asked her, 'What name have you given the child this time?'

"She answered, 'Middling.'

"When they had worked again for a bit, Mrs. Maria once more pretended she heard some one calling, and cried, 'All right! all right! I'm coming!'

"'Well; who's calling you again?' asked Mr. Nicholas.

---

³ Campbell's 'Popular Tales of the West Highlands,' No. 65.
⁴ Hahn, 'Griechische Märchen,' No. 89.

"'Oh! I have again to stand godmother.'

"'Hem! it's strange, though, that they're so constantly inviting you!'

"'That's because they're so fond of me, Mr. Nicholas!'

"'Go, then: but mind you come back soon.'

"So Mrs. Maria stole off to the honey a third time, ate it and the bread clean up, tipped the pot over, and then returned to Mr. Nicholas.

"'What name, pray, have you given *this* child?' he asked.

"'Tipling,' was Mrs. Maria's answer.

"So they continued for a while to dig: but at last Mr. Nicholas said, 'Shall we have our dinner now or later?'

"'Let us dine now,' said Mrs. Maria.

"So Mr. Nicholas went to the bush where the honey and the loaves were concealed, and found the pot tipped over. At that he was furious, and, running to Mrs. Maria, cried, 'You have eaten all the white bread and the honey, and in return I will now eat you.'

"'No; I declare I have not eaten them, Mr. Nicholas! You must have been mistaken in what you saw; do go again, and look more carefully!'

"'Well, I will,' he said: 'but if I find nothing, rest assured there's something in store for you.'

"While Mr. Nicholas was gone to look again, Mrs. Maria ran off and hid herself in a hole. When the wolf was returning, he saw her just creeping in: so he took a hooked stick to get her out with. But when he hooked a root, the fox cried out, 'Oh! oh! my poor leg! oh! oh! my poor leg!' When he caught hold of her foot, she made game of him: 'Pull at the root, you ass! Yes; pull at the root, you ass!' And so it went on for a long time, till at last the wolf got tired and went off; and so Mrs. Maria escaped from Mr. Nicholas."

The search for Popular Tales in various countries, after it was once fairly started, proceeded apace; and we now possess collections from Germany, Norway, Scotland, Wallachia, Servia, Lithuania, Albania, Greece,[5] and other

---

[5] The titles of these are: Grimm's 'Deutsche Kinder und Haus-Märchen;' Dasent's 'Popular Tales from the Norse;' Campbell's 'Popular Tales of the West Highlands;' Schott's 'Walachische Märchen;' Wuk's 'Serbische Märchen;' Schleicher's 'Litauische Märchen;' Von Hahn's 'Griechische und Albanesische Märchen.' I have made use of the prefaces and notes to most of these collections, especially those of Von Hahn. His comparative table of the tales of various countries is invaluable.

parts of Europe; those from Greece, with which we are most nearly concerned, being the latest additions to the growing family. France and Italy are represented by older collections, in which, though the character of the tales has been altered, and their setting elaborately wrought to suit the taste of fashionable readers, yet the outline and the original traits have mostly been preserved. In particular, the Neapolitan collection, which was published by Basile more than two centuries ago under the title of 'Il Pentamerone,' notwithstanding the redundancy of fanciful wit with which it is tricked out, retains here and there points of a very primitive character, which have been lost elsewhere, and has a value of its own, on account of the early date at which it was brought together. With these European stories, again, have been compared those found in Asia, such as the Armenian tales,[6] the ancient Indian collection called the Five Books, or 'Pantschatantra,' and the 'Arabian Nights.' With these materials before us we have ample facilities for a scientific investigation of the subject; and the result of the comparison is the proof of the most striking correspondences between the stories of all countries, such as can only be accounted for by assuming their common origin. That so surprising an idea, and one involving so many difficulties, was not at once accepted, is only what we might have expected; at first it was attempted to show that the resemblances were accidental, and great stress was laid on the tendency of the human mind to produce similar images on the recurrence of similar circumstances: but it very soon appeared that resistance was hopeless, and that the resemblances, amounting in some cases to absolute identity, were too strong

---

[6] In Haxthausen's 'Transcaucasia.'

and too numerous to be accounted for in any other way.

Take, for instance, the story of Cinderella, one that certainly is sufficiently peculiar, and not likely to have been invented more than once. On opening Von Hahn's collection of Modern Greek tales, we are almost startled at finding this as the second story, with all the details to which we are accustomed—the youngest of three sisters, who is ill-treated and neglected; the present of three dresses of surpassing beauty, which she receives from some beneficent power, in return for an act of piety towards her deceased mother; her appearance in these on three separate occasions, unexpected and unrecognised, either at church or at a ball; her winning the prince's love, and three times escaping from him; her dropping one slipper on the last occasion, and being subsequently recognised in her mean attire in consequence of its fitting her and her only; her marriage with the prince, and the disappointment of her sisters. The same story is found, with slight variations, among the German, the Servian, and the Neapolitan[7] tales, in some of which points are developed and worked out which are only hinted at in

[7] The following address of the Prince to the slipper, after Cinderella has escaped from him, is a favourable specimen of the graceful banter by which the stories are accompanied in Basile's version :—" The King took the slipper into his hand and said, 'If the foundation is so beautiful, what must be the appearance of the house? Thou beautiful candlestick, on which was placed the light that enkindles me; stand of the beautiful kettle, in which my life is boiling; beautiful cork, attached to Cupid's fishing-line, with which he has caught my soul; lo! thus I embrace thee and press thee to my heart; and though I may not reach the tree, yet do I adore the roots; though I may not have the capital, yet do I kiss the pedestal! Hitherto hast thou been the prison of a snowy foot, now thou art the fetter of an unhappy heart; thou didst elevate by an inch and-a-half the empress of my life, and to that degree is increased by thy means the joy of my life, as long as I possess thee and retain thee.' "—Basile's ' Pentamerone,' in Liebrecht's German translation, i. p. 87.

others. In the Servian and Greek versions the story is introduced by a marvellous narrative of the mother being killed and eaten, and the youngest daughter rewarded for collecting, burying, and watching her bones: in the Servian tale the mother is first changed into a cow.

Or take the story of the Snake-child, which is found in the Greek, the Albanian, and several other European collections, and also in the Indian tales. In this a childless woman prays to God that she may have some offspring, even if it were a snake. In the course of time her wish is fulfilled to the letter, and a snake is born to her: nevertheless, she rears it as her child, and it grows up to maturity. At last the snake requests his mother to get him a wife; and though most of her neighbours laugh at the idea, when she proposes it, a step-mother is found willing to part with her step-daughter on these terms. At first the maiden is sorely distressed at her strange fate, but on visiting her mother's grave she is instructed in a dream that her future husband is not really a snake, but a handsome youth concealed under that form. After marriage he appears to her by night in his human shape, laying aside his serpent's skin, to which he returns by day. On this his wife devises the following plan. On the succeeding night, when her husband is asleep, she takes his serpent's skin, and hands it out to her mother-in-law, who puts it into the fire and burns it; after which he permanently retains his human form. With this, which is the Modern Greek version of the story, the Indian tale in the 'Pantschatantra,'[8] called "The Brahmin's Enchanted Son," almost identically corresponds, though there the snake does not speak or show signs of intelligence until

---

[8] Benfey's 'Pantschatantra,' ii. pp. 144 foll. Compare also Frere's 'Old Deccan Days,' p. 193.

after marriage. The similar European stories have evidently come from the same original, but present curious variations in the form under which the youth is born. In the Wallachian tale it is a gourd; in one of the German tales an ass's foal, in another a hedgehog; in the Albanian and Neapolitan it is a snake, but in the latter it is not born, but adopted as the woman's child—evidently a later and weaker form of the story. Other remarkable peculiarities occur in various parts. In almost all the versions the husband is conscious of his cast-off skin being burnt. In the Servian tale the wife pours cold water over his body, while it is going on, to save his life. In several of the others he is forced by the loss of his original shape to leave his wife, who is unable to bring forth her child until she recovers him; and the history of her wanderings and search forms a continuation of the story. She finds three enchantresses, who are called in the Albanian tale the "Sisters of the Sun;" in the Servian the "Mother of the Sun," "Mother of the Moon," and "Mother of the Wind;" in the Wallachian, "Mother Wednesday," "Mother Friday," and "Mother Sunday;" in the Neapolitan they are spoken of simply as aged women; in one of the German tales it is the sun, moon, and wind themselves that are consulted. From them she receives valuable magic gifts, and the last whom she visits directs her to the residence of her husband, who has married another wife. By means of the gifts she persuades her rival to allow her to meet her husband, whom she thus reclaims and carries off.

The stories which have now been described are only specimens taken from among a great number, in which the resemblances are equally striking; and there is a still larger class which correspond less completely, but yet in a manner that cannot be accounted for by accidental coincidence. To illustrate this, let us take two passages, one from the north, the other from the south of Europe. The following occurs in Mr. Dasent's 'Popular Tales from the Norse':[9] it describes a boy's escape from a Troll on an enchanted horse by means of three objects which he is instructed to carry with him.

"So when the lad had got on the horse, off they went at such a rate he couldn't at all tell how they went. But when he had ridden awhile, the horse said, 'I think I hear a noise; look round. Can you see anything?'

"'Yes, there are ever so many coming after us; at least a score,' said the lad.

"'Aye, aye; that's the Troll coming,' said the horse; 'now he's after us with his pack.'

"So they rode on awhile, until those who followed were close behind them.

"'Now throw your bramble-bush rod behind you, over your shoulder,' said the horse; 'but mind you throw it a good way off my back.'

"So the lad did that, and all at once a close, thick bramble-wood grew up behind them. So the lad rode on a long, long time, while the Troll and his crew had to go home to fetch something to hew their way through the wood. But at last the horse said again, 'Look behind you; can you see anything now?'

"'Yes, ever so many,' said the lad; 'as many as would fill a large church.'

"'Aye, aye; that's the Troll and his crew,' said the horse, 'now he's got more to back him. But now throw down the stone, and mind you throw it far behind me.'

"As soon as the lad did what the horse said, up rose a great black hill of rock behind him. So the Troll had to be off home to fetch something to mine his way through the rock, and while the Troll did that the lad rode a good bit farther on. But still the horse begged him to look behind him, and then he saw a troop like a whole army behind him, and they glistened in the sunbeams.

"'Aye, aye,' said the horse, 'that's the Troll, and now he's got his whole band with him; so throw the pitcher of water behind you, but mind you don't spill any of it upon me.'

[9] P. 362.

"So the lad did that; but in spite of all the pains he took he still spilt one drop on the horse's flank. So it became a great deep lake; and because of that one drop the horse found himself far out in it, but still he swam safe to land. But when the Trolls came to the lake they lay down to drink it dry, and so they swilled and swilled till they burst."

Now compare this with a passage in Von Hahn's 'Greek Tales,'[10] where a boy and girl are represented as flying from their unnatural mother, of whose desire to kill them and eat their flesh they are warned by the house dog.

"'What are we to take with us?' inquired the boy of Pulja.

"'What are we to take with us, say you? I know not, Asterinos,' the girl answered; 'stop, though! take a knife, a comb, and a handful of salt.'

"So they took this, and the dog too, and started off and ran for a bit; and while they were running they saw from a distance their mother, who was pursuing them.

"Then said Asterinos to his sister, 'Look! there's our mother running after us, and she'll soon be up with us.'

"'Run, love, run!' the girl replied; 'she won't catch us.'

"'She's upon us now, Pulja, dear!'

"'Throw the knife behind you.'

"The boy did so, and out of it there arose an immense plain between them and their mother; but she ran more quickly than the children, and again came near to them.

"'She's catching us now!' cried the boy again.

"'Run, love, run; she won't come up with us.'

"'Here she is!'

"'Throw the comb behind you.'

"He did so, and out of it grew a thick, thick wood. Still the mother worked her way through the thicket also, and when she reached the children the third time they threw the salt behind them, and that became a sea and their mother could not pass through. The children stood still on the shore, and looked across."

---

[10] Vol. i. p. 66. The same thing is found in Carleton's 'Stories of the Irish Peasantry.' i. p. 37, and in a somewhat mutilated form in Frere's 'Old Deccan Days,' p. 69.

The idea which forms the groundwork of both these narratives is evidently the same.

Even in respect of the form into which the narrative is cast, the household stories present a family likeness to one another. This is seen, for instance, in the repetition of an action or event three times (as is usually the case), before the desired effect is obtained. And in most of the collections we find peculiar short rhyming verses, frequently of an enigmatical character, embedded in the stories—a feature which has suggested the idea that at one period they may have existed in verse throughout. But the conclusion which is drawn from the comparison is irresistible, that the popular tales of all these various countries, notwithstanding the distance by which they are separated from one another, are to be referred to a common source, and possess a very high antiquity. The result of the investigation fully justifies the satisfaction with which William Grimm, in one of the later editions (1850) of his German Stories, remarks on the fulfilment of his views, and the gradual acceptance of the theory which he originally started. "How solitary," he exclaims, "was the position of our collection, when first it came to light, and what a rich crop has sprung up since that time! People used then to smile indulgently at the assertion, that thoughts and views of things were here preserved, the beginnings of which went back into the darkness of antiquity; now it is hardly ever contradicted. In the search for these stories their scientific value is recognised, and care is taken not to tamper with their contents, whereas formerly they were regarded merely as the valueless products of a playful fancy, which must be content with whatever treatment they might receive."[11]

---

[11] 'Kinder und Haus-Märchen,' iii. p. 360.

The question then arises, if we admit the fact of their common origin, how are we to account for their dispersion? At what period, and by what agency, were they carried in different directions to the widely separate localities in which we now discover them? In reply to this, two opposing theories have been started: according to the one, they were the common inheritance of all the races of the Indo-European family, and, perhaps, in some instances, of the whole human race: according to the other, they were of Indian extraction, and were imported from that country into other parts of Asia and Europe, where they have subsequently taken root.

The latter of these two views receives great additional weight from being maintained by Dr. Benfey, whose very extensive knowledge of Sanscrit literature, and acquaintance with the popular literature of Oriental and other countries, enables him to speak with great authority on the subject. According to him, there is found in India a vast number of stories which may have served as the originals of those that exist elsewhere, and by means of them it is possible to explain the occurrence in European tales of features which seem to have had their origin in Buddhism, such as contrasts of the gratitude of animals with the ingratitude of men, and more especially the attribution of beneficent qualities to the snake. Whenever traces of these stories are found in Europe earlier than the tenth century A.D., as, for instance, in some of the Greek myths, and elsewhere in classical literature, he believes that their introduction is due to occasional intercourse between the East and the West, and that they were orally transmitted by merchants, travellers, and others, who communicated them to those with whom they were brought into contact. But from the time that the Mahometans began to invade and conquer India in the tenth century, the acquaintance of Europe with Asia was continually on the increase, and resulted, among other things, in the translation of the Indian stories into western languages. The 'Pantschatantra' was translated into Persian, Arabic, Hebrew, Greek, and Latin, and its contents, owing to their striking and attractive character, spread with great rapidity over the continent of Europe.[12] This view he maintains with a vast array of learning, and with great critical acuteness. But it fails to account for the facts. Not only are there numerous stories which find no counterpart in the Indian tales, but it is impossible to explain in this way the extraordinary profusion in which this popular literature is found over the whole face of Europe. Turn the soil where you will, and a plentiful crop at once springs up in unbounded luxuriance and variety. And besides this, from the way in which the germ and salient point of one story is introduced as a supplementary incident into others, either in the same or in different countries, and from the frequent interchanges that are found, parts of the same story being united or separated in various collections, and combined from time to time with other stories, we seem to infer a native growth, and a higher antiquity than this explanation implies. We must remember also how scanty were the means of communication between countries all through the Middle Ages, and in how small a degree the lower classes have migrated, even to the present day, especially the women, in whose possession these treasures are chiefly found. To this must be added the barrier of language, which would form an almost insurmountable obstacle in the way of their transmission from one people to another, and the deeply-rooted feeling of shyness already noticed,

---

[12] Benfey's 'Pantschatantra,' i. Pref. p. xxii. foll.; 5 foll.

which is everywhere found to hinder the possessors from imparting them to others.

Of course it is quite possible that in some cases tales may have been adopted from Indian and other sources into this literature of the peasants; but the more closely we examine each particular instance, the less reason there usually seems to be for admitting such a supposition. Thus, to take the instances on which Benfey is disposed to lay most stress—the contrast of the gratitude of animals with the ingratitude of men, and good qualities being imputed to the snake—from the prominence of these ideas in Buddhist tenets, and their occurrence in the 'Pantschatantra,' there seems a considerable probability in favour of their having been imported; but it is not impossible that both may have been original Aryan notions, and that the latter may have had a mythological significance. It is remarkable, also, that the story of the "Three Grateful Animals" should be found in so many countries, if it is of comparatively late introduction. In connection with this idea of the stories being borrowed, the Modern Greek tales have a peculiar interest, because their country has been exposed to external influences in an unusual degree. Thus, the Greeks, from residing in the chief Turkish cities, are constantly thrown in the way of eastern story-tellers, and they possess, and are fond of reading, a translation of the 'Arabian Nights' in their own language. Yet the result of a careful comparison is to show that, though there are many points in the Greek tales which might appear to be derived from this source,—such as the slave of the ring, the forty chambers of which only one is forbidden to be entered, the death of the forty thieves by oil being poured on them, and others,—yet the resemblances are equally numerous between the 'Arabian Nights' and the German tales. Again, from the large infusion of Slavonic blood into the Greek race, and the presence of numerous Wallachs among them, who in part speak their language, we might have expected that there would be signs of their having received contributions from these peoples; but here, too, though the coincidences are very striking between the Modern Greek and the Servian and Wallachian tales, they are not more so than those which occur elsewhere. Indeed (and this is especially worthy of notice), in several instances the Greek stories correspond with those of Lithuania, where there is no trace of any similar ones in the intervening countries, or, in fact, in any of the collections hitherto made. Nor have the Italians, despite their residence in Greece, and commercial intercourse with that country, left any proofs of their influence in this respect: wherever the same story is common to both, the Greek retains the more primitive form. The same is the case with the ancient Greek literature. Here, again, the points of resemblance that are found in the tales are striking: the maiden bound to a rock and doomed to be devoured by a monster, reminds us of Andromeda; the flute which makes all men, woods, and mountains that hear it to dance, recals Orpheus; the choice of Achilles, the fortunes of Danae, and other well-known stories also appear. But most of these are equally found in other countries; and on the whole there is as much trace of Greek myths in the German stories as in those of Modern Greece.

We may here notice a point which at first sight is somewhat perplexing, namely, the difficulty of conceiving that these stories, with their peculiar element of the marvellous, could have underlain the ancient Greek mythology, and existed side by side with it. This, however, we must suppose, if we believe them to have been inherited by the inhabitants from primitive ages. That children's stories did exist among the Greeks and Romans is shown by several passages in classical writers; Quinctilian speaks of them in connection with Æsop's fables, and it is likely enough that they were of the same character as those which Tertullian describes as being told at Carthage in his day about "enchantresses' towers and sun-combs"[13]—an expression which would just describe some of the features of the modern tales. But the marvellous element disappears with the advance of civilization from epic poetry and other forms of cultivated literature, and this would sufficiently account for their not having appeared above the surface, or left more than occasional traces of their influence. Anyhow we have a tolerably exact parallel in modern times, as they have underlain not only Christianity, but all the hagiology of the Middle Ages, and still maintain themselves, and influence the lower forms of popular belief. In parts of Devonshire at the present day, and probably elsewhere in England, it is customary, if a child is ill, to slit the branch of a sapling tree, pass the child through, and tie it up again; after which, if the tree lives, the child will recover, but thenceforward its life is bound up with and dependent on the life of the tree. What is this but another form of the idea which gave birth to the hamadryad of Greek mythology? And what was the source of both of them but the popular tales and their corresponding mythology, in which the same idea occurs in both these forms?

We are driven, then, to the conclusion that these tales are not Indian, but Indo-European; that they, or at

---

[13] *Lamiae turres et pectines solis.* See the refs. in Grimm's 'Märchen,' iii. pp. 273, 274.

least the embryo forms out of which they were developed, were the common possession of the Aryan races before they separated from one another, and were carried by them in different directions to the countries which they ultimately occupied. Many of them, no doubt, have been amplified or modified in the course of ages; but, notwithstanding this, they may be regarded as embodying and pourtraying in a very remarkable manner the character, the forms of belief, and the modes of thought, of the earliest representatives of our branch of mankind. They form an important constituent in what Schott has expressively called "the intellectual band which, in spite of war and hate, indissolubly binds together the European peoples, and perhaps the human race." Whether, indeed, they are universal, and not confined to any one family of nations, it is impossible as yet to decide. From the 'Arabian Nights' we can draw no argument on this subject, as there can be little doubt that they, or at least large portions of them, have been drawn from Indo-European sources. But when we find, not only in other parts of Asia, but even among remote African tribes, stories that correspond to those with which we are acquainted, such as those which Mr. Dasent has given in the excellent preface to his translation of the Norse tales, we can hardly escape from the conclusion that they are a common tradition derived from the one original stock. In all probability, however, the number of these will prove to be small relatively to the entire aggregate which we possess, and the ideas which they embody to be for the most part simple. The older the human race is proved to be (and its great antiquity seems to be in a fair way of being demonstrated by several converging lines of evidence), the greater will be the difficulty of supposing that any large number of tales, or any that possess considerable intricacy, have been handed down from the earliest ages.

When we consider the prospect which this view of the popular tales opens out before us, we can hardly over-estimate the interest attaching to them. While investigators and antiquarians are labouring with admirable perseverance to reconstruct the history of ruined kingdoms, in these we possess ready to our hand the monuments of a period of far higher antiquity, and illustrating subjects more closely connected with men's inner life; written, no doubt, in cypher in some cases, but yet not more difficult to interpret than cuneiform or hieroglyphic inscriptions. To this we must add the truly poetical spirit which breathes throughout them, a feature the value of which Schiller was not slow to recognise. It is related of him that during his last illness, when his creative genius was thrown into a state of feverish activity by the process of designing new productions, he exclaimed, "Give me popular stories and tales of chivalry; in them are found the materials of all that is good and beautiful."[14]

Having thus carried back these stories to a common origin, we are naturally led to enquire further whether any account can be given of the process by which they were formed. Were they at that early period the ingenious creations of the imagination, or is it possible to trace their development in any more scientific manner? From all that we know of similar products of the mind, the probabilities are strong against any explanation which would refer them to accident, and the family likeness which runs through them seems to suggest that some law must have presided over their formation.

---

[14] See Schott, 'Walachische Märchen,' p. 317.

Without implying that in every instance the same cause can necessarily be assigned, we can confidently affirm that in a large number of cases they are the lingering sounds of world-old myths, the latest stage into which they have passed in the course of their development. To make use of William Grimm's comparison, "the mythical element in them resembles tiny fragments of a broken gem, which lie dispersed upon the ground, where it is overgrown with grass and flowers, and require a penetrating eye to discover them."[15] That this was the case, Walter Scott, from his intimate acquaintance with stories and romances of all kinds, had already in part divined, although the question had hardly been mooted in his time. "A work of great interest," he says, "might be compiled upon the origin of popular fiction, and the transmission of similar tales from age to age and from country to country. The mythology of one period would then appear to pass into the romance of the next century, and that into the nursery tale of the subsequent ages."[16] One of the best instances by which we can illustrate this is a story that is well known to all of us, "The Sleeping Beauty," which exists in most countries in the form with which we are familiar, and is found also, altered and almost travestied, as the basis of many other stories.

The tale commences, it will be remembered, with a feast on the occasion of the Princess's birth, to which among others the wise women of the kingdom were invited, in order that they might be propitious towards the infant. They were thirteen in number, but one of them was omitted in the invitation, and determined to revenge herself for the slight. Accordingly, during the

[15] Grimm, 'Märchen,' iii. p. 409.
[16] Appendix to the 'Lady of the Lake,' note 3 D.

feast, when her sisters were bestowing on the young Princess all the blessings which they had at their disposal, she appeared in the midst of them, and declared that at fifteen years of age she should die by the prick of a spindle. Eleven of the others had already spoken in their turn, but one remained to speak, and as she could not reverse the curse of her malignant sister, she did what she could to mitigate it by substituting for death a sleep of a hundred years' duration. In consequence of this, the father did everything that was in his power to prevent his daughter from touching a spindle, but all in vain: when the appointed time arrived, an old woman appeared at the castle with a spindle, and with this the Princess pricked her finger, and at once fell into a deep sleep. At the same moment every person and thing within the castle became motionless by enchantment, while a thorn hedge grew up all round it, and prevented access to it from every side. At the expiration of the hundred years comes the *denouement*, which Tennyson has described in his poem. The Prince who is fated to break the spell appears, and before him the hedge, which had resisted everyone else, gives way, and by his kiss the Princess, and with her the whole enchanted household, awaken from their long slumber.

For the prototype of this tale we must look to the story of "Brynhildr and Sigurd" in the ancient Volsung tale, the original of the "Nibelungen Lied" of Germany. Brynhildr, a Valkyrie, or Houri of the Northern mythology, in consequence of her having given victory in battle to the wrong hero, is pricked by Odin in the head with his thorn, on which she sinks into a deep enchanted sleep. The remote castle in which she lies is surrounded by a wall of flame, and in accordance with the will of the weird sisters, the Fates of the North, she is doomed to slumber there until a man should come bold enough to ride through the flame and take her for his bride. The destined hero is Sigurd, who, mounted on his horse Gran, makes his way boldly through the enveloping fire, and wakes her from her sleep.

In passing from this shape into the story of the Sleeping Beauty, this narrative follows the ordinary law of the development of a myth in its later stages, viz., that all the characters and objects in it are more and more materialised, and reduced to the level of ordinary life. The Fates assume the form of wise women; the Valkyrie appears as the king's daughter; Odin and his thorn of sleep are the old woman and her spindle; and the wall of flame becomes the impassable hedge of thorn.

We may now proceed to trace the mythological significance of the early story, and through it of the form it has subsequently taken; again remembering that the process of the formation of myths, as well as their later development, is a realistic one, the phenomena and powers of nature being personified, and their motions expressed in the form of human actions. It is the story of summer and winter. The thorn of Odin, which produces sleep by its touch, is the influence of the inclement winter. By means of this, at the season appointed by the Fates, Brynhildr, the goddess of flowers, who represents the beauty of nature, is thrown into a death-like slumber. While in this state she is encompassed and guarded by flames, the symbols of the unapproachable world below, and so of the extinction of life in all nature. Sigurd is the sun-god, who approaches in the spring time, and by his loving touch restores life to the world.[17]

---

[17] The mythical character of this story is now pretty generally recognised. Thus, in the latest edition of the 'Nibelungen Lied,' Dr. Braunfels remarks

The Neapolitan version of the story, though in some respects less complete than the others, for it omits all mention of the thorn hedge, contains some features of a remarkably primitive character. The Prince is represented as approaching the castle, when he is out hunting, and the falcon which is perched on his wrist flies off and enters a window of the building, and refuses to return. This point recalls Sigurd's hawk in the Volsung tale, which flies into Brynhildr's tower, as the hero is drawing near. Again, the cause of the Princess's sleep is a prick, not from the spindle, but from a fibre of flax, which penetrates under her finger-nail and remains there; and her restoration to consciousness is brought about in the following manner. She is described as bearing two children to the Prince while still asleep, which are tended by fairies, and are called Sun and Moon; and on one occasion one of the infants, being unable to reach her breast, sucks her finger instead, and in doing so withdraws the flax, the cause of her slumber, from under her nail, on which she wakes. The meaning of this can hardly be mistaken, as describing the deliverance of the earth from the power of winter by means of the heavenly bodies. In the corresponding French tale, "La belle au bois dormant," the children are called by the equally significant names of Dawn and Day.

It is quite surprising how often, and in how many shapes, the idea of a maiden condemned to an enchanted

---

in his introduction (p. v.):—"The story of Siegfried (Sigurd) and Brünhild is derived from the earliest times of the German race; perhaps from those days in which the Germans, Greeks, and Indians, had not yet branched off into independent peoples, but were living united in their original relationship on the plateaux of Asia. This accounts for its being discovered, though variously worked up, among all the different families of nations of the so-called Caucasian race. Its proper meaning is everywhere of a mythical character."

sleep, which forms the germ of this story, reappears in the Popular Tales. Thus in the story called "Snowy-white," which occurs in the German, Wallachian, and Albanian collections, the fair one who bears this name falls into this state from wearing a hairpin, a comb, or a ring (the two former recall the thorn in Brynhildr's head), which is conveyed to her by the artifices of a wicked step-mother. In each case she wakes as soon as the cause of her enchantment is removed, which is effected by means of a stranger prince. In another very curious Wallachian story, called "The Devil in the Spigot," the heroine refuses to marry any one who cannot tire her out in dancing. When everyone else has failed, at last the devil himself comes, and succeeds; on which he turns the princess, the court, the castle, and all that it contains, into stone. In this condition they are destined to remain, until some one appears who can outwit him. At last a "jolly companion" comes by, who in the course of a drinking bout persuades the devil to get inside a cask, and then turns the cock upon him. The spell is at once broken, and the return of everything to its former condition is very beautifully described. The whole thing sounds like a burlesque of the Sleeping Beauty, and it is still more strange to conceive of it as akin to the story of Brynhildr; but there can hardly be any doubt that it is so, and that it, and numberless others, describe the paralysis of nature in winter, and the return of life in spring.[18]

If it were in our power to decipher the meaning of the majority of these stories, we should probably find that the ideas from which they sprang were very limited in number. Another large class may be referred to the

---

[18] The story is also found in India; see Frere's 'Old Deccan Days,' pp. 91, 92.

ideas which in Greek mythology are embodied in the fable of Danae. This, it will be remembered, consists of two parts, the confinement of Danae in a tower or a subterraneous chamber, where she is visited by Zeus in a shower of gold, and gives birth to Perseus; and her exposure with her child in a chest on the sea. In the popular tales these were found independent of one another, and are continually recurring. The former of them represents the birth of day, Perseus being the day-god, and the golden shower the rays of golden light. The latter is not so easy to explain, but it probably refers to the return of the sun from west to east between sunset and sunrise, a phenomenon always perplexing to the ancients, and which some of the Greek poets describe as taking place in a golden bowl.[19] It would carry us too far if we were to attempt to examine the evidence on which these statements rest, or to refer to the various collections of stories for examples.[20] But there is no difficulty in this constant repetition of the same notions in various forms. If we are to adopt the theory of mythology of which Professor Max Müller is the representative in England,[21] and suppose that myths were origi-

---

[19] Preller, 'Griechische Mythologie,' i. pp. 339, 340.

[20] An argument, the force of which depends entirely on an induction based on careful observation, necessarily suffers when only one or two instances are produced. But hardly any one who has studied the popular tales carefully has failed to discover that there is a mythical element in them; and this, when analysed by the help of mythology, has generally appeared to resolve itself in great measure into an embodiment of the forces and phases of nature. The process is not one of guess-work.

[21] This may be briefly stated as follows. At a certain period of human development large classes of ideas were expressed in figurative language by means of personification. In the course of time these expressions were misunderstood, and they became myths, the persons being conceived of as real persons, and their acts as real acts: these myths were afterwards further corrupted by being supplemented for purposes of explanation.

Thus among the early races the sun was spoken of as a powerful hero;

nally nothing more than figurative language, which, when its significance was lost, became crystallized into the strange and beautiful forms that have come down to us, it is only what we should expect, if the ideas which they represent are very limited. The persons who employed this language were a race of shepherds, living before the dawn of civilization, and whose thoughts must consequently have been centered on a few objects, which they conceived with intense vividness, and represented to themselves under a great variety of forms. In an especial manner, the contrast of summer and winter must have come home to them with a force which we who live in warm dwellings, with ample means of amusement and occupation, can hardly understand. In speaking of the Montenegrins, I have mentioned how it was described to me in the country that during the winter months they have absolutely nothing to do except to try to keep themselves warm. This represents the normal condition of mountaineers during that season. With what intensity of longing, then, must they look forward to the return of spring, with all its sources of enjoyment and forms of beauty, and how natural it seems that they should

---

the bright days were his herd of cows—350 in number, as they are described in the 'Odyssey' (xii. 127, *seq.*), thereby representing approximately the days of the year; the winter was a wild, mis-shapen, furious giant, dwelling in the West—the land of sunset, and so of darkness; the pale light and short days of winter are described as the concealing of the cows by the giant in his cavern—an idea found in the Vedas. Accordingly the return of spring would be expressed in the following way—the hero brings back the cows from the cave of the giant in the West. When the original significance is lost, this appears as the Greek myth of Hercules and Geryon. Hercules (the Sun-God) carries off the oxen (the long bright days) from the cave of the monster Geryon (their concealment during stormy winter) at Gades (in the region of darkness). This simple outline is afterwards filled in by numerous additional facts, to explain the reason for Hercules' visit to the West, and to help out the story in other ways.

constantly have it in their thoughts and on their lips, and describe it with a lavish profusion of images! Nor need we be surprised if the shapes which these mythical ideas assume sometimes imply an observation of the aspects of nature, such as we might expect rather from a modern painter than from an Aryan shepherd.[22] Even at the present day those of the lower classes, to whom the changes in the atmosphere are a matter of primary importance, the peasant and the sailor, possess a knowledge of the forms of clouds, and a sympathy with nature's moods, of which we have but a faint conception.[23] And that ideas of this kind were familiar to those early races is proved by the evidence of the Vedas, which were the literature of a people in that stage of development, for in them striking descriptions of natural objects are to be found.

The mythical origin which has thus been assigned to these stories may seem in part to explain their extraordinary permanence, and the slight degree in which they have been affected by influences of time and place. They had, in fact, at one time a religious character; their kings and queens were understood to represent the old gods and goddesses, a feeling of which some trace appears still to remain, as their place is occasionally supplied by Christ and the Virgin. In this way a feeling of awe would grow up round them, and this may have had no slight influence in producing the secresy in which they are now maintained. But it is not necessary to suppose that in all cases the stories have passed through the par-

---

[22] Such, for instance, as the *cirrus* clouds, which, as Mr. Baring-Gould points out ('Book of Were-Wolves,' p. 176. Compare 'Curious Myths,' second series, pp. 297 foll.), are represented by the feather-dresses worn by certain fairy beings in most of the collections of popular tales.

[23] *See* this point interestingly worked out in an essay in the 'Fortnightly Review' (No. V. p. 590), on "The Clouds and the Poor."

ticular form which we call a myth; on the contrary, not unfrequently it would seem that the material out of which both myths and stories were formed was condensed at once into the latter more concrete shape, for otherwise we could not account for the very exact correspondences that are found in those possessed by widely distant peoples.[24] In some instances the myth and the corresponding story may have existed side by side. Nor does it follow that when a story that has been derived from a myth is common to many races, it has always been inherited by them in the same manner; for in one case the story itself may have been common to the ancestors of those races, while in another they may only have possessed the myth, which after their separation was corrupted into stories more or less similar. It may, perhaps, be possible at some future time, when the subject has been more thoroughly sifted, and more complete collections have been made, to discover the relative ages of many of the stories, by observing the character and amount of their correspondences, and by noticing the races to which they are or are not common, as well as by tracing the more or less primitive features which they contain. In this way some may be shown to have belonged to the earliest representatives of the human race, some to the Aryan family and to it only, some to the Teutonic nations, and so on in other cases.

It will have been inferred from what has been already said that the modern Greek tales do not materially differ from those of other countries. At the same time they

---

[24] It is easy to see how, from such figurative language as "rosy smiles and pearly tears," a story might arise about a maiden, from whose mouth a rose dropped whenever she laughed, and whose tears fell in the shape of pearls (Hahn, No. 28). The same thing must have taken place in instances much less easy of solution.

possess certain peculiarities, some of which it may be well to notice. Their style of composition, like that of most other collections, has distinguishing features of its own. Thus, while the narrative of the Lithuanian tales is heavy and matter-of-fact, that of the German simple, homely, and humorous, and that of the Wallachian poetical, but lengthy and diffuse, the Greek stories are clear, bright, lively, and full of point. In this respect, indeed, they are scarcely inferior to those of any collection, though it is evident that they have hardly at all been tampered with by civilisation, unlike the Neapolitan stories, which have been elaborately wrought up by art. Again, in respect of their contents, though here, as elsewhere, the marvellous element, which is essential to popular tales, for the most part excludes any notice of the existing state of the country, and the salient points are always fabulous incidents, yet, on the whole, there is more than usual reference to modern times. Thus, the evil and designing character is often embodied in a Jew or dervish, the one representing the hereditary objects of Greek antipathy, the other the most fanatical of Mussulmans: while the monk, when he appears, is a good and beneficent personage. So also a *hadji* is introduced, who has made a pilgrimage to the Holy Land; the prodigal son betakes himself to Athens and Corfu, to spend his substance in riotous living; a lady, whose husband has been taken from her by enchantment, has a suit of Frankish clothes made, in which to go abroad in search of him; and when she arrives in India, and in the character of a physician cures the king's son of an illness, she asks in return that all the chiefs of India may be summoned to a banquet, which the king replies, is a very small matter for him. Again, a variety of customs, both ecclesiastical and secular, are introduced in different places: the use of incense; the killing and eating the Easter lamb; the practice of collecting and throwing into a common heap the bones of the dead some time after burial; offering coffee and sweetmeats to a guest on his arrival; inflating a lamb before skinning, to loosen the skin; the bridegroom's "best man" at a wedding stealing a spoon from the table in the bride's house; *pobratim*, or contracting fraternal friendships. Of peculiar creations of the imagination which are not found in the stories of other countries, there are not many; according to Von Hahn, only three—the Dog's-head, a monster in man's shape, that eats human flesh; the Half-man, a being with only half a head, half a nose, half a mouth, half a body, one hand and one foot; and the Karakisa, a female monster, who also devours human beings, and whose strength is derived from bathing, and lies in the foam on her lips.

But by far the most remarkable being that we meet with here—though it is not peculiar to this collection, for its counterpart is found elsewhere—is the Drakos. This name merely signifies "dragon," for δράκος and δράκοντας in modern Greek are only other forms of δράκων, in the same way as γέρος, γέροντας, are of γέρων; but it may be well to retain it, in order to mark the distinctive features of that which it represents. The Drakos is a huge, ponderous creature in human form, who likewise delights in man's flesh, and is endowed with enormous strength, and proportionately small intellect. Most of the stories into which he is introduced as a principal character turn on his violence and awkwardness, and the ease with which he is outwitted by ingenious men. In all these respects he resembles the German and English Giant, and the Troll of the Norse tales. Like them, also, he is the reputed artificer of ancient works which seem to have required great force for their accomplishment; and just as in the British islands we have a Giant's Causeway and Giant's Cromlech, so in the island of Carystos in the Ægean the prostrate Hellenic columns in the neighbourhood of the city are said to have been flung down from above by the Drakos; and in Tenos, a smooth rock, which descends precipitously into the sea, is called the Dragoness's Washing-board, from its resemblance to the places where Greek women wash their clothes. The Dragoness, who is the wife of the Drakos, and is generally represented as his female counterpart, sometimes also bears the name of Lamia; and as this name suggests a connection with ancient Greek mythology, and additional peculiarities are now and then attributed to her, it may be well to reserve her for further consideration in the next chapter.

It will be seen from this that though this being bears the name of dragon, he is entirely devoid of the characteristics of that imaginary creature. In all the ordinary occurrences of life he acts like a man. He fetches wood and water, goes out hunting, and even goes to church, notwithstanding his questionable character, leaving his dragoness at home meanwhile to kill and cook the guest who is staying with them. He is principally distinguished by his clumsiness in everything that he has to perform, and his ignorance of the simplest arts of civilised life. In one of the stories the dragoness is conciliated by a man who shows her how to bake properly, as up to that time she had been accustomed to throw pieces of dough on to the live coals. Similarly in the Norse tales, as Mr. Dasent observes, the Trolls can neither brew nor wash properly, and agriculture is a secret of mankind which the Giants are eager to learn, but which is a branch of knowledge beyond their power to attain.[25] But though in all this there is no trace of the dragon nature, yet a comparison of corresponding conceptions in other countries shows that it must originally have existed. Thus, while the Lithuanian and Servian stories agree with the Greek in having eliminated all but the human characteristics, the Wallachian tales in the same connection introduce real dragons, which breathe flame and pestilential vapours, and are provided with wings and claws. Perhaps the idea of their being able to assume human form may have underlain the conception of them from the first; but the name of dragon, which remains in all cases, though the form may have been lost, proves that that shape was originally common to all.

The following story may serve to illustrate the characteristics of the Drakos of which I have spoken:—

"Mr. Lazarus and the Dragons.[26]

"There was once a cobbler, called Lazarus, who was a great lover of honey. One day, when he was eating some at his work, such a number of flies collected that he killed forty of them at a blow. On this, off he went, and had a sword made, on which was written, 'With one blow I have killed forty.' As soon as the sword was ready he started and went into the world, and when he was two days' distance from his home he came to a spring of water, by the side of which he laid himself down and went to sleep.

"Now it happened that that spot was inhabited by dragons; and when one of them came to the spring to fetch water he found Lazarus sleeping there, and read what was written on his sword. On this he went off to his comrades, and related to them what he had seen, and they advised him to conclude a fraternal friendship (*pobratim*) with the powerful stranger.

"Accordingly, the Drakos returned to the spring, waked Lazarus,

---

[25] Dasent's 'Popular Tales from the Norse,' Introd., p. cxl.
[26] Hahn, 'Griechische Märchen,' No. 23.

and proposed to him that, if he was agreeable, they should make a fraternal friendship between them.

"Lazarus replied that he was agreeable; and after a priest had pronounced a blessing on their fraternal friendship, he went with him to the other dragons and abode with them. After some days, they told him it was their custom to go in turn to fetch wood and water, and as he was now one of their company, he must also take his turn. So they went out first for wood and water; but at last Lazarus's turn also came to fetch water. The dragons, however, had a jack that contained fifty gallons of water, and when it was empty Lazarus had great difficulty in dragging it to the spring; and as he could not have carried it when it was full, he did not fill it, but dug up the earth round about the spring instead.

"When they saw that Lazarus stayed out so long, they sent one of their party after him to see what was become of him; and when he came to the spring, Lazarus said to him, 'Don't let us bother ourselves any more with fetching water from day to day. I'll carry the entire spring home, and so we shall be free from the trouble.' 'Body and soul!' cried the Drakos, 'don't do that, Mr. Lazarus, or we shall all die of thirst. We'd much rather fetch water in turn and let you go free.'

"Next came Lazarus's turn to fetch wood. The dragons that went for wood used always to take a whole tree on their shoulders and carry it home; and as he could not follow their example, he went into the wood, tied all the trees to one another with the rope they used for carrying, and remained in the wood till evening. Again the dragons sent one of their number after him to see what was become of him; and when he asked what he was intending to do, Lazarus answered, 'I am going to drag away the whole wood at once, that we may have rest.' Then cried the Drakos, 'Not for the world, Mr. Lazarus, else we shall die of cold; we'd rather go for wood ourselves, and let you remain free.' So the Drakos pulled up a tree, took it upon his shoulder, and carried it home.

"When they had lived some time in this way, the dragons became tired of Lazarus, and devised among them a plan for killing him. It was arranged that each of them should give him a blow with an axe during the night, when he was asleep. Lazarus, however, had overheard this; so in the evening he took a log of wood, covered it with his cloak, put it in his sleeping-place, and then concealed himself. During the night the dragons came, and each of them gave the log a blow with the axe till it was broken in pieces. Then, thinking they had attained their object, they lay down again. After this, Lazarus took up the log and threw it out, and lay down. Towards daybreak he began to groan, and when the dragons heard him they asked what was the matter with him; whereupon he answered, 'The gnats have been stinging me dreadfully.' Then the dragons were afraid, because they thought Lazarus took their blows with the axe for gnat-stings, and determined to get rid of him at any price. Accordingly, the next morning, they inquired of him whether he had a wife and children; and whether he would not like some time to pay them a visit, as they would give him a sack full of gold to take with him. This he agreed to, only demanding in addition that one of the dragons should carry the gold for him to his home. So one of them went with him, and carried the sack. When they came near to Lazarus's house, he said to the Drakos, 'Wait here a minute, for I must go and tie up my children that they may not eat you up.' So he went and tied his children with thick cords, and said to them, 'As soon as the Drakos comes in sight, cry with all your might, "Dragon's flesh! Dragon's flesh!"' Accordingly, when the Drakos approached, the children screamed, 'Dragon's flesh! Dragon's flesh!' and at that the Drakos was so terrified that he let fall the sack and ran off. On the way a she fox met him, and asked him why he looked so frightened? and he answered that he was afraid of Mr. Lazarus's children, who had been within an ace of eating him. Then the fox exclaimed, laughing, 'What! were you afraid of Mr. Lazarus's children? Why, he had two fowls, and one of them I devoured for him yesterday, and the other I am going for this moment. If you won't believe it, come along with me, and you shall see it; only you must tie yourself on to my tail.' So the Drakos tied himself to her tail, and returned with her to Lazarus's house, to see what she'd be up to. There, however, stood Lazarus in ambush, with his gun; and when he saw the fox coming with the Drakos he cried out to her, 'Didn't I tell you to bring me all the Dragons? and now you are bringing me one only?' When the Drakos heard that, he turned right round, took to his heels, and ran so quickly that the fox was dashed to pieces on the stones. But when Mr. Lazarus had got free from the dragons he built himself a magnificent house with their gold, and passed the rest of his life in splendour and enjoyment."

In the German story, which corresponds to the first half of this one (Grimm., No. 20), the cobbler's antagonists are giants; the number of flies that he kills is seven; and the fate of the slain is inscribed on a girdle. The three trials of strength and skill by which he establishes his influence over the giants are, first, squeezing water out of a cream-cheese, which the giant takes for a white stone; secondly, letting off a bird that he has caught into the air, which the giant supposes to be a stone, and thinks it has been thrown so high that it can never come down again; thirdly, undertaking to carry one end of a felled tree, if the giant will carry the other, and then climbing up into the branches behind his back, and making him carry him. The incident of the blows of the axe on the bed is also given, but independently, and occurs also in our Jack the Giant Killer. In one of the Norse tales (Dasent, pp. 41 foll.) a boy gets the mastery over a Troll by three similar trials. The first of these is squeezing water out of the cheese, like the German story; the second, offering to carry home the spring of water, like the Greek; the third is an eating match, in which the boy drops his share of the porridge into a bag, and at last persuades the Troll to rip himself up, like our Jack the Giant Killer. The latter part of the tale of Mr. Lazarus is found in an Indian story in the 'Pantschatantra' (Benfey, i. p. 506), where the Drakos is replaced by a tiger, which is frightened away by a mother telling her children not to dispute about eating a tiger a-piece, but to divide this one between them, and then look out for another. A jackal persuades the tiger to return with him, on which the mother again sends him off headlong by abusing the jackal for only bringing one tiger, when she had ordered him to bring three.

If we proceed to compare the Drakos with beings which are interchangeable with him in other collections of stories, we shall arrive at some remarkable results, which throw a curious light on the question of their mythological character. We have already seen that these Greek dragons were once real dragons, and that among the Wallachians this original form has been retained. We have seen also that the same things are attributed to them as are attributed in the north of Europe to giants and Trolls. But besides these, there are other conceptions which correspond to them elsewhere. In the Albanian story of "Snowy-white" (Hahn, No. 103), the persecuted maiden, when flying from her stepmother, after wandering into a wilderness, seeks for shelter in a house inhabited by dragons of the Greek type. In the corresponding German story (Grimm, No. 53), the occupants of the place are dwarfs, a class of beings which in the Northern mythology possess many characteristics in common with their larger brethren, the giants.[27] In the Wallachian story (Schott, No. 5) the conception has been weakened, and they appear simply as a band of robbers. Another supernatural figure that often appears in the Greek stories is a Moor, that is, a black man, who appears and disappears with equal suddenness, and in some cases can be summoned by licking an enchanted ring. From this last trait among others we perceive that he is the "slave of the ring" of the 'Arabian Nights.' Again, in speaking of the ideas the Albanians entertain with regard to hidden treasures, I have mentioned that the powers which are supposed to guard them are negroes and snakes; and in some of the tales also there are points which seem to imply a relationship between these two. Thus in one of the Greek tales (Hahn, No. 110) a Moor, by spitting

---

[27] See Grimm, 'Deutsche Mythologie,' i. p. 519. The same writer remarks that in the German mythology the giant, as the sensual element, is believed to have come first in the order of creation; the dwarf, who represents wit and craft, comes next; and last comes man, who is the intermediate point between the two (ibid., p. 485).

into a boy's mouth, confers on him the power of obtaining the fulfilment of any wish that he expresses. In the Servian tales (Wuk, No. 3), the King of the Serpents, by the same act of spitting into the mouth, enables a shepherd to understand the language of beasts.[24]

The correspondences which thus admit of these conceptions being interchanged with one another, imply that there is something which they all possess in common; and that common element is the fact that they, or at least the originals from which they spring, are powers of the nether world. Such a figure as a supernatural black man requires no explanation. Snakes and dragons again are naturally associated with the powers of evil. In the case of the Trolls, the same thing is proved by their being unable under any circumstances to bear the sight of the sun; if once his light falls upon them, they burst. And in the same way we may explain the strange custom in Greece, that the new-born child before baptism bears the name of Drakos; the reason apparently being that at that time it is believed to be especially exposed to the influence of malignant powers. Besides this nether-world character, or perhaps as a cause of it, we trace in the giants, Trolls, and dragons, evident signs of their having represented great natural forces, such as the frost, the wind, and the tempest, which are gradually being subdued by the advancing civilisation of man. Their old order, like that of the Titans of Greek mythology, was that of a more chaotic period, which is destined to pass away. An old woman of the isle of Andros, a teller of stories, related to Von Hahn that when men first came to Andros to the dragons, there was living

---

[24] In the tales of the West Highlands, a man obtains the knowledge of all secrets by tasting the broth in which a white snake was boiled.—'Campbell's Tales,' No. xlvii.

there a very old Drakos, who was blind. He begged that a man might be brought to him for him to feel, and by that means to form an idea of what he was; in order, however, to preserve the man from harm, a ploughshare was laid upon his head. The old Drakos seized the ploughshare, and crushed it to dust.[29] Mr. Dasent relates a similar story. "See what pretty playthings, mother!" cries the giant's daughter, as she unties her apron, and shews her a plough, and horses, and peasant. "Back with them this instant," cries the mother in wrath, "and put them down as carefully as you can, for these playthings can do our race great harm, and when these come we must budge."[30]

If it be enquired further, which of the powers of nature in particular the Drakos embodies, we may answer with some certainty that it is the thunderstorm. Mr. Baring-Gould, in his 'Book of Were-wolves,'[31] has brought forward some very remarkable legends of different periods, in which the fury and the ravages of a dragon are described in such a way as to leave no doubt on the mind that it is a violent storm which is intended to be pourtrayed. And he observes generally that "the dragon of popular mythology is nothing else than the thunderstorm, rising at the horizon, rushing with expanded, winnowing, black pennons across the sky, darting out its forked fiery tongue, and belching fire." In modern Greece it is in the Lamia rather than her husband that the clearest traces of this can be discovered, for she is regarded by the people as presiding over the whirlwind and waterspout, and as being especially unpropitious to sailors. But with regard to the Drakos also there is evidence of the same thing in the Greek tales. Not

---

[29] Hahn, 'Griechische Märchen,' i. p. 39.
[30] 'Popular Tales from the Norse,' p. cxl.   [31] Pp. 172 foll.

only does a Drakos in one place (ii. p. 270) change himself into the form of a cloud, but in another (ii. p. 50) he is described as coming with a crashing sound, and destroying the fruit on the trees. This passage is a very remarkable one. A tree that bears golden apples is robbed of them as soon as they are ripe. The eldest son of the king, to whom it belongs, takes up his position by night under the tree to watch for the thief. "As he was standing there," the story proceeds, "all at once the earth began to quake, and a cloud descended on the apple-tree in the midst of fearful thunder and lightning, and from out of it was reached forth, as it were, a hand; and lo! an apple was gone." The same thing takes place the next night, when the second son keeps watch; but both of them are too much frightened to make any resistance. When the youngest's turn comes, who, as usual, is the most successful, he also "hears a terrific crash, and sees a black cloud, which stretches out towards the last remaining apple." He, however, retains his presence of mind, and fires into the midst of the cloud, which immediately disappears, leaving the apple untouched. The next day he tracks the blood-marks, which lead him to a stone with an iron ring in a mountain side; and when he has pulled this up, a scalding-hot steam arises from the bowels of the earth. This proves to be the breath of the Drakos, and when the youth has descended he finds the thief, and ultimately kills him. In this story the Drakos appears not only as a being of the nether world, but also in the form of a thunderstorm.

Let us conclude with one of the ballads, the subject of which is this same monster. Here again he is frightened away by blustering words. From the very elliptical way in which the story is told it would appear either that the audience are supposed to be familiar with it, or that the recitation must be accompanied with some kind of pantomime.

"JACK AND THE DRAKOS.

"Who was it passed this way last night? who was it passed by singing?
He waked i' their nests the nightingales, and from the rocks the spirits;
He wakened, too, a dragoness, who by her lord was sleeping.
The Drakos was exceeding wroth, great was his indignation:
'Who was the man that dared to sing? By all the gods I'll eat him!'
(Jack is caught by the Drakos.)
'O mercy, Drako! let me go; leave me for five days longer:
For Sunday is my wedding-day, and Monday is my fête-day;
And Wednesday morning to my home I lead my lovely lady.'
The sun went down, the night came on, the moon had waned in heaven,
And now the glittering morning-star was on the point of setting.
(Jack and his bride make their appearance at the house of the Drakos.)
'Ha! welcome! here my breakfast comes; welcome! here comes my supper!'
'Your breakfast, sir, shall be on stones, you shall have rocks for supper;
For know that I'm the lightning's son, the daughter of the thunder.'
Away! be off, I pray you, Jack! you and your bride together.'[32]

---

[32] Passow, 'Pop. Carm.,' No. 509.

CHAPTER XXX.

ON THE CLASSICAL SUPERSTITIONS EXISTING AMONG THE MODERN GREEKS.

Question as to the Descent of the Modern Greeks — Evidence derived from Superstitions — The Nereids — Their Beauty and Malevolent Disposition — Ballad on the Nereids — Propensity to steal Children — Origin of the Superstition — Elements or Genii — Spirits residing in Wells — The Lamia — Gello — The Fates and Furies — State of the Dead — Hades — Charon — His enlarged Functions — Ballad of "The Maiden in Hades" — Comparison of the Ancient and Modern Conceptions — "Passage Money" — Feeding the Manes — Other Ancient Customs — Christian Use of Pagan Fables.

THE question of the origin and race of the modern Greeks has been warmly discussed of late years, especially in Germany. Until some time after the end of the War of Independence they were believed to be lineal descendants of the ancient Hellenes, and no small portion of the enthusiasm which that struggle excited in Western Europe was caused by the sentiments which their supposed classical descent inspired. At last, however, Professor Fallmerayer of Munich broached the theory that they were in no wise related to the old Greeks, but of Slavonic origin, the descendants of those hordes who had poured over the country in successive waves of immigration during the Middle Ages. The controversy that followed took the same course, and passed through the same stages, as many similar controversies. At first the suggestion was ridiculed by some as a paradox, and decried as striking at the root of all Phil-hellenism, while others were enlisted in its favour by the attraction of novelty. A third party, approaching the subject in a more impartial spirit, though unwilling to accept so disappointing a conclusion, were yet so startled by the array of facts and ingenious arguments presented to them, as to have an uncomfortable presentiment that it might turn out to be true. Then came the process of more complete investigation. On the one side there was the undeniable fact, that a great proportion of the names of places throughout Greece are to this day Slavonic, showing how permanent must have been the influence of that race, and how wide their dispersion throughout the country. There were also strong passages from the Byzantine writers, describing the depopulation of the country, and its complete occupation by barbarian Slavonians. Nor could it be denied that, except in some of the islands, and in other remote places, like the southern part of the Morea, the distinctive Greek physiognomy was no longer to be found. On the other hand, it was pointed out that notwithstanding all the differences which these changes implied, the language, the strongest of all tests of national identity, had remained the same, allowing only for the changes which are naturally wrought by time; and that it was almost wholly free from any traces of Slavonic influence. Great stress was laid on the extraordinary resemblance of character between the ancient and modern Greeks, and that too in points presenting the sharpest contrast to that of the Slavonic races,— their quickness, restlessness, and versatility, accompanied by a corresponding want of perseverance; their intelligence, ingenuity, and thirst for knowledge; their unbounded love of talking and argument; their elasticity of temperament, fertility of imagination, and irrepressible self-confidence, giving birth to an inordinate ambition their personal vanity and selfishness, combined with a strong feeling of patriotism, and resulting in political mistrust, party spirit and intrigue; their fondness for trickery and sharp practice. Together with these general features, a multitude of minor traits were adduced more easy to observe than to describe, but implying a remarkable similarity in modes of thought. Lastly, by careful investigation a large number of customs and superstitions were brought to light, which, though here and there altered or obscured, retained unmistakeably the impress of classical times. The result of the discussion seems to be the re-establishment of the old belief, only subject to very considerable modifications. No one can doubt that the physical element in the Greek race is very largely derived from Slavonic and other extraneous sources; but at the same time the Hellenic blood appears to have retained through the lapse of ages that same power of assimilation, by which in ancient times it amalgamated with itself the large Pelasgic population of the country. In this way, though physically the modern Greeks may have but a slight, perhaps a very slight, claim to call the ancient Greeks their forefathers, yet in all that really constitutes a people, their character, feelings, and ideas, they are their lineal descendants. This conclusion is now pretty generally received, and is approved, among others, by Mr. Finlay, whose severe impartiality adds weight to his authority.

The particular point which we are now about to consider, viz., the Pagan superstitions and forms of belief which have been inherited by the modern Greeks from classical times, is one of considerable interest, and at the same time requires no little care in its treatment. Both in this and in the customs which appear to be Hellenic, it is easy to be led away by slight resemblances, which a more careful inspection will prove to be accidental, and thus to run the risk of tracing a connection between ancient and modern times, where in reality none exists. It is necessary, also, in some cases, where the modern conception appears to be compounded of several elements, to distinguish as far as possible what is evidently classical from that which belongs to an independent system of mythology, and to observe the changes which, from whatever cause, have been introduced into it at a later period. But, after all this has been done, a sufficient residuum will be found to show that the old beliefs have not wholly died out, but are still at work in a variety of ways in the minds of the people. With the greater divinities we must not expect to meet, as they were driven off the scene by the advance of Christianity; though many Christian rites still bear witness to the severity of the struggle, by the remains of Pagan observances which lie embedded in them. It is rather the minor deities, and those associated with man's ordinary life, that have escaped the brunt of the storm, and retired to live in a dim twilight of popular belief. The consequence of this is, that these superstitions have a strange fragmentary character, as being disintegrated from what was once a complete system, and surprise us all the more when we meet them, from our inability to explain them, or discover the exact foundation on which they rest.

Let us commence with the Nereids, the belief in whom is universal throughout the continent of Greece, the islands of the Archipelago, and the shores of Asia Minor. The name is found in a variety of forms—νεράϊδες, ναράϊδες, and with the prefix a, ἀναράϊδες, ἀνεράδες, ἀνεράγδαις; and reminds us that their classical prototypes, together with their father Nereus, derived their name from a root connected with the modern Greek word for "water," νερό. They are also known as "the good ladies" (καλαὶς κυράδες, καλαὶς ἀρχόντισσαις), though this term is applied to them, not in the same way as the Northern fairies are called "the good people," from an idea that they are generally friendly to man, but as an euphemism, for, as we shall presently see, they are almost universally malevolent. They dwell by the sea shore (νεράϊδα τοῦ γιαλοῦ), and have been seen, like the Nereids of old, dancing by moonlight on the surface of the sea; but this is not their only home, for they are equally found by rivers and fountains, in caverns and woods, and on the mountains; so that they may be regarded as representing generally the nymphs of antiquity. By day they are rarely seen, though a peasant in Argolis described how he had often seen one of them, with her green hair ornamented with pearls and corals, drying her dripping garments on the cliffs:[1] under these circumstances they are wont to wreak their vengeance on those who chance to see them; and a story is told of a woman who was beaten by them for disturbing them in the forest at a meal, and was afterwards struck dead when she turned back to watch them.[2] By night they appear, and, like all the nymphs of ancient times, amuse themselves with graceful dances; but even then their influence is dangerous, and the unlucky wight by whom they are seen in crossing a river is doomed to speedy death, unless a priest counteracts the evil influence by reading verses from the Scripture. This part of the superstition Mr. Newton refers to the fact that it is on the banks of rivers by night that the most dangerous fevers are caught.[3] They are clad in white, and are not diminutive in stature, like our fairies, nor do they in other respects differ from beautiful women. Amongst the Mainotes, however, in the south of Laconia,

---

[1] Wachsmuth, 'Das alte Griechenland im Neuen,' p. 32.
[2] Hahn, 'Märchen,' p. 80.   [3] 'Discoveries in the Levant,' i. p. 211.

the conception of them is somewhat different, and some of the peculiarities attributed to them seem to recal the ancient satyrs. They are there described as "three maidens of the most exquisite beauty, except that they have goats' legs and feet, who dance continually round the summit of Mount Scardamyla."[4] The number three which is here assigned to them is also occasionally found in the Popular Tales; but usually no fixed number is spoken of, though they are seldom seen in large companies, like the fairies.

The beauty of the Nereids is proverbial. In the Romaic ballads, when the lover desires to extol his mistress's charms, he can find no higher compliment than to compare her to a Nereid.[5] But notwithstanding their personal appearance, their nature is full of malice, and continually prompts them to work fatal mischief. In another of the ballads an obdurate lady, who is charged with ruining her lover, is addressed as "godchild of Charon, and offspring of the Nereid.[6] In the Popular Tales they are called the Devil's daughters,[7] and are for ever engaged in spiteful acts. Accordingly, some of the most harmful natural phenomena, such as the whirlwind and the storm, are ascribed to their agency; when the winds are sweeping everything furiously before them, they are supposed to be present, and presiding over the work of destruction.[8] At such times the only way of appeasing their fury is to crouch down while they are passing over; and a story is told of a girl who, on re-

---

[4] Fauriel, 'Chants Populaires,' Pref. p. lxxxv.
[5] Passow, 'Carm. Pop.,' Distich No. 692.
[6] —— τοῦ Χάρου συντεκνάδι,
τῆς Ἀνεράγδας γέννημα.
—Passow, 'Pop. Carm.' Distich, No. 653.
[7] Hahn, No. 54.
[8] Leo Allatius, 'De quorundam Græcorum Opinationibus,' p. 160.

fusing to do so, was carried off by them, and kept a prisoner in the mountains.⁹ At Athens during a whirlwind old women are accustomed to murmur a sort of incantation to soothe them, using the words, "Milk and honey in your path."¹⁰ Nor is the expression a meaningless one, for it is an ascertained fact that below the hill of the Museum at Athens, at Thebes, and in other places, honeycakes and other gifts are from time to time offered to them.¹¹

These beings are supposed to possess the power of bewitching young persons, whom they bring under their influence, and ultimately cause to pine away. As an instance of this, the following story is related by Professor Ross as having come under his own cognisance in the village of Chalandri, near Athens, in the year 1833. Seeing the wife of the priest of the place in mourning, he enquired the cause of it, on which she gave him the following reply. "I had a daughter," she said, "a girl of from twelve to thirteen years of age, who gave proof of a very peculiar temperament. Though we all behaved kindly to her, she was continually melancholy in her disposition, and as often as it was in her power would escape from the village to the wooded slopes of the mountain (Brilessus). There she used to wander about in solitude all the day long, early in the morning as well as late in the evening: sometimes she even took off her upper garment, and tied round her only a small light dress, in order to be less impeded in running and jumping. We did not venture to prevent her, for we were well aware that the Nereids had ensnared her; but we were deeply distressed. It was of no avail that my husband frequently took her into the church and recited

⁹ Hahn, No. 81. ¹⁰ Wachsmuth, p. 31.
¹¹ Ross, 'Inselreisen,' iii. p. 182.

prayers over her; the Holy Virgin had no power to help her. After she had gone on in this way for a considerable time, she fell into a still deeper melancholy, and at last died not long ago. When we buried her the neighbours said: 'Do not be surprised at her death, for the Nereids desired to have her; only two days ago we saw her dancing in the midst of them.'"[12]

In their other relations to men the same malevolent disposition appears. When they meet with mortal lovers, indeed, and become attached to them—a thing which is said not unfrequently to happen, and was related also of the nymphs of ancient times—they bestow on them rich gifts, and raise them to a condition of opulence. In the island of Casos, Professor Ross was informed by a Greek in whose house he stayed, that his uncle, who possessed more than a thousand head of cattle, was believed to have been indebted to a Nereid for this wealth; and when at a later period he fell into poverty and misfortune, this was ascribed to her anger at his unfaithfulness.[13] In almost all other instances where men are reported to have had connection with them, similar injurious consequences have ensued. Accordingly, when a young man is suddenly afflicted with illness, especially such as affects the nerves and results in any kind of paralysis, he is believed to have fallen under their displeasure. Similarly in classical mythology a hamadryad who was attached to a mortal lover, when on one occasion he slighted a communication which she sent him by a bee, her accustomed messenger, vented her indignation on him by rendering him lame.[14] It is especially at midday that they exercise this noxious

---

[12] Ross, iii. p. 181.
[13] Ibid., iii. p. 45. Comp. Leo Allatius, p. 160.
[14] Donaldson's 'Pindar,' p. 386.

power; at that time they are accustomed to rest in the shade of trees, especially of the plane and poplar, and by the side of springs and running water, and those who pass by and disturb them are apt to receive a stroke.[15] In like manner in ancient times Pan was supposed to rest from his hunting at midday, and to be tired and out of temper, so that the shepherds were afraid to sing or play the flute at that hour.[16] But in such cases the modern Greeks shrink from mentioning directly the influence from which the malady is supposed to proceed, and use an euphemistic form, such as "he has received it from without," or "his time had arrived." In the Popular Tales they are usually described as marrying men who succeed in stealing from them their feather dresses, or, in some instances, their wings, which they put off when bathing: in these their power resides, and as long as they are kept out of their reach they are prisoners: but if ever they succeed in regaining them, they at once make their escape, and cannot be persuaded to return. In one story, a Nereid lives for some time with a man, in all respects like other women, and bears him a son; but one day, when dancing was going on, she was reminded of her former dances in the air, and longed to regain her wings to use them for that purpose; her husband is at last persuaded to restore them to her, on her promising faithfully not to fly away; but when she gets possession of them, she flies three times round the dancing-place, and then disappears. Afterwards she was wont to return by day in her husband's absence, and attend to the household and take care of the child; but when her husband met her and reproached her, she exclaimed,— "This is how the Nereids deceive you."[17]

[15] Leo Allatius, p. 159; Wachsmuth, p. 30.
[16] Theocrit., 'Idyll,' i. 15-18.   [17] Hahn, No. 83.

The following song from the island of Carpathos illustrates several of the points which have just been mentioned:—

"Nine thousand sheep composed the flock; nine brothers had to watch them:
Five of them went to win a kiss, and three to see their sweethearts;
Yani alone was left behind among the sheep to watch them.
And so he watched the sheep with care, the flock he duly guarded.
Oft had his mother said to him, full often had she charged him,
'Beware, my Yani, oh! beware; I warn thee, my good Yani;
Avoid a solitary tree, descend not to the meadows,
And by the upland river-bank play not upon the reed-pipe,
For thither will assemble all the Nereids of the river.'
But he regarded not her words and disobeyed her warning.
He sought a solitary tree, he went down to the meadows,
And by the upland river-bank he played upon the reed-pipe,
And thither were assembled all the Nereids of the river:
'Play to us, Yani! play to us; oh! play upon the reed-pipe.
Wilt thou have money? lo! 'tis thine. Wilt thou have costly jewels?
Wilt thou the fairest of the fair, the loveliest of maidens?'
'Give me no gold, I seek it not; give me no costly jewels;
Nor yet the fairest of the fair, the loveliest of maidens.
Thee, my Eudocia, I desire: thee, amiable songstress;
Who singest so at break of day that all the birds awaken.'"[18]

And so we may suppose, like Goethe's mermaid—

"Sie sprach zu ihm, sie sang zu ihm;
Da war's um ihn geschehn."

But the reason for which, far more than for any other, the Nereids are dreaded, is their propensity to carry off children. So firmly is this believed, that when a birth takes place, Mr. Newton tells us, no person whatever is allowed to enter the house, except the midwife, till the child has been blessed by a priest; and it is customary for forty days after the birth to close the house-door

[18] Passow, 'Carm. Pop.,' No. 525.

at sunset, and never to open it after that hour for fear the *aneradcs* should enter and carry off the child.[19] For the same reason, women with child wear a variety of amulets to counteract their influence, and must especially avoid lying down under a plane-tree or poplar, or in the neighbourhood of a spring or running water.[20] Some of the infants which they thus carry off are said to be restored by them after a time, heightened in beauty and enriched with handsome gifts; but others they retain as their own.[21] By doing so, however, they are supposed to incur the wrath of Heaven, and it is described how one of them was struck with lightning in consequence of a child being thus detained.[22] The same practice seems to some extent to have been attributed to the Nymphs in old times, for in many epitaphs on children that died at an early age, they are spoken of as having been carried off by the Nymphs.[23] Sometimes, also, the Nereids exchange one of their own young ones for a human child; but the changeling which is thus left behind is said never to live.[24] This trick is one of the mischievous freaks of the northern fairies.

From the account that has now been given it will be seen, I think, that the Nereid of the modern Greeks is a compound idea, derived from more than one source, and representing partly the classical nymphs, and partly a set of beings more or less corresponding to our fairies: with the former of these they are identified even by the peasants at the present day, for when Ross, in climbing over some precipitous places, found some cavities in the rocks, and an inscription referring to them, con-

---

[19] Newton, i. p. 211.
[20] Wachsmuth, p. 71.
[21] Leo Allatius, p. 158.
[22] Hahn, 'Märchen,' No. 84.
[23] Preller, 'Griechische Mythologie,' i. p. 565, *note*.
[24] Pashley's 'Crete,' ii. p. 216.

taining an invocation of the Nymphs, the shepherds who accompanied him exclaimed—"Ah! those are the beings that we now call Nereids."[25] That they have lost the genial and friendly disposition which the Greeks of old used to attribute to these deities, and have assumed a demon character, is probably to be accounted for by their having been outlawed by Christianity, and forced to take up a position of hostility. But in the places they frequent, and in most of their characteristics, they are still the same; so that when Milton, in describing the overthrow of Paganism, said—

> "From haunted spring and dale,
> Edged with poplar pale,
>   The parting genius is with sighing sent;
> With flower-inwoven tresses torn,
> The nymphs in twilight shade of tangled thickets mourn"—

he fixed on one of the most essential elements in Greek mythology, but at the same time had hardly realised, perhaps, how permanent and ineradicable this belief was.

As to the other constituents in the modern conception, that which corresponds to the fairies, it can hardly be supposed to have been borrowed from northern nations, both on account of the distance which separates them, and the marked points of contrast between the two beliefs, which are not less striking than the similarities. Thus the fairies are of both sexes: they are diminutive in stature; they are frequently benevolent to men; in respect of their demon nature they are divided into different orders—fairies, brownies, and black elves or dwarfs, according to the proportion in which they possess it; they also form a *people* or community to them-

---

[25] Ross, iii. p. 45.

selves. In the Nereids these characteristics are wholly wanting. Nor do they appear to be more closely allied to the *vila*, or Servian fairy, the belief in which might more easily have been naturalised among the modern Greeks; though the fact of the Nereids being found in Crete and Asia Minor would in itself be an objection to assigning them a Slavonic origin, since the Greeks of these countries have been but little Slavonised. It is rather to the earlier mythology, which is found in the Popular Tales, that the correspondences between these conceptions of various races are to be referred; and there are some features of the Nereids which can certainly be referred to this source. The feather dresses which they are sometimes described as wearing, are distinctly an Indo-European idea, and connect them with the *apsaras* of the Veda, which symbolise the floating clouds of the upper sky. This cloud-origin may perhaps be traced where the Nereids themselves are said to assume the form of clouds.[26]

There is, however, another class of supernatural beings among the Modern Greeks, in which the ancient pervading spirit of polytheism is still more distinctly seen. As among the Hellenes of old every fountain, every well, every tree had its presiding spirit, not only haunting it, but existing in it, and identified with it, so among their descendants the same tendency is visible, and has peopled the whole land with innumerable genii, whose watchful care is supposed to protect the objects to which they are attached. These in the modern language are called στοιχεῖα, or *elements*, a term that was formerly applied by the Platonists to the spirits which were believed to exist in the earth, air, fire, and water, and afterwards passed into an

---

[26] Hahn, 'Märchen,' i. p. 299.

appellation for demons in general.[27] They are all of the male sex, and are believed to reside in rocks, caverns, and trees; in rivers, wells, and fountains; and also in houses. In the latter the form that they assume is generally that of a dragon or snake, which may remind us of the "guardian serpent" by which the houses of the Greeks and Romans were haunted of old.[28] This domestic genius is supposed to be a cause of prosperity to the household, though, if any injury is offered to it, it is sure to entail some harm on the inmates; in consequence of this it is greatly revered, and almost worshipped. Those that reside in other places are usually innocuous, but not always so, for one spirit, a sort of hamadryad, which inhabits the olive-tree and the walnut, is said to be very fatal to the lives of men.[29] Another strange legend relates that the genius of the sea was at war with the genius of the plane-tree for a thousand years, and that when a struggle took place between them, and one was conquered, numerous deaths took place in the neighbourhood.[30] The subject of one of the ballads also is the death of a hunter in consequence of his killing the "en-

---

[27] See Ducange, s. v. στοιχεῖον. Sophocles, in his 'Dictionary of later and Byzantine Greek,' attributes the later meaning of the word to a misinterpretation of the passages where St. Paul speaks of the στοιχεῖα τοῦ κόσμου (Col. i. 8, 20; Gal. iv. 3, 9). In process of time, he thinks, the ignorant imagined that *evil spirits* or *demons* were meant. It seems more probable, however, that the modern signification arose as has been suggested above, and that this use of the word caused the misinterpretation of those texts. In a curious passage of one of the modern ballads, three brothers are represented as watching their flocks together with οἱ τρεῖς στοιχειὰ τοῦ κόσμου.—Passow, No. 524. I strongly suspect that here the underlying idea is that of the Holy Trinity.

[28] Herod., viii. 41; Soph. Phil., 1328; Livy, i. 56; Virg. 'Æn.,' v. 95. Amongst the Greeks, as far as I know, we have only evidence of its having haunted the temples; but the epithet, οἰκουρός, which is applied to it, seems to imply that amongst them also it was found in houses.

[29] Passow, No. 514.   [30] Ibid., Index, p. 634.

chanted stag" (τὸ στοιχειωμένο ἐλάφι), which is described as having his horns in the form of a cross, a star on his forehead, and a figure of the Virgin between his shoulders.[31] This song is found in Euboea, and therefore in curious proximity to the spot where Agamemnon incurred the wrath of Artemis by killing her sacred deer. At the same time the story bears some resemblance to the western legend of S. Hubert, who was converted when out hunting by the apparition of a deer with the cross between his horns; and it is just possible that this may have passed into these countries at the time of the Crusades.

Speaking of "the demons that are met with in the forests," Mr. Newton tells us—"I asked a peasant what they were like. He said that he believed them to have μὲ συμπάθεια σᾶς (the equivalent of *con rispetto parlato*) goats' legs and tails, and said they were like the figures painted on Greek vases. He admitted, however, that he had never seen one himself." Mr. Newton remarks that these seem to be a tradition of the old Greek satyrs.[32] More numerous are the stories relating to those that make their abode in wells. These are also called τελώνια —a name which at one time was applied by sailors to the appearance called St. Elmo's light.[33] In the island of Myconos it is customary, before drawing water from a well, to offer three salutations to the genius that resides there.[34] These beings frequently appear under the form of a diminutive black man, and are said to have enticed girls to descend with them into their wells, where they have passed through suites of splendid and handsomely furnished apartments beneath the earth.[35] The same

---

[31] Passow, No. 516.   [32] Newton, i. p. 211.   [33] Wachsmuth, p. 58.
[34] Villoison, quoted by Pashley, i. p. 93.   [35] Leo Allatius, p. 166.

writer who mentions this tells us also of a well in a grotto at Chios, to which a peculiar superstition is attached. From this, at midnight, a man issues, seated on a spirited charger, and accompanied by a rushing sound; and afterwards he returns in the same manner, and disappears with his horse into the well. The man is called Benia, and persons who drink of the water of the well are said to lose their senses; so that it is customary in the island, when laughing at any one for his folly, to say, "you have drunk of the fountain of Benia."

The following highly imaginative story is related by Mr. Pashley, as having been told him by an inhabitant of Sphakia, in Crete. Its character, however, is hardly classical:—

"Two men went, on a fine moonlight night, up the lofty mountains, intending to hunt the agrimia (game). They heard a great tumult, and at first supposed it to be caused by people coming to obtain snow, to take it into the city; but, as they drew nearer, they heard the sound of musical instruments and varied sports. The men soon discovered that these were not mortals, but an assemblage of demoniacal beings, all of whom were clothed in variegated garments, and rode on horses, some of which were white, and others of different colours. It appeared that there were both men and women, on foot and on horseback, a multitude of people; and the men were white as doves, and the women beautiful as the sunbeams. It was also evident that they were carrying something like a bier. The mountaineers determined to shoot at the aerial host, as they passed on singing,—

'We go, we go, to fetch the lady-bride
From the steep rock, a solitary nymph.'

As soon as the shot was fired, those who were last in the procession exclaimed, 'They've murdered our bridegroom! they've murdered our bridegroom!' and as they made this exclamation they wept, and shrieked, and fled." [36]

Another being, whose name and some of whose attri-

---

[36] Pashley, ii. p. 217, 218.

butes have been perpetuated in the modern mythology,
is the Lamia. In ancient times this monster was conceived of in two different ways. According to the earlier
belief she was the mother of Scylla, and had once been
beautiful and beloved by Zeus; but when Hera in envy
robbed her of her children, she retired to a dark cavern
in the midst of precipitous rocks, where she became
hideous and misshapen, and delighted in carrying off and
devouring other women's offspring. The later story is
that which Keats has embodied in his poem, of an en-
chantress in the form of a serpent, who had the power of
assuming the appearance of a beautiful woman. It is
the former rather than the latter of these which the
modern conception resembles. She is now, as has been
mentioned in connection with the Popular Tales, the wife
of the Drakos, and is characterised by the same want of
intellect which distinguishes that monster: she also delights in human flesh, and kills children, so that when a
child dies suddenly, the people say, "it was throttled by the
Lamia." In consequence of this the name is used, as it was
in ancient times, to frighten disobedient children. Besides
this, in some parts of Greece she is regarded as a spirit
of the sea (λάμια τοῦ πελάγου), and the queen of the
Nereids, in which character she presides over storms, and
in particular over the waterspout and whirlwind: in consequence of this, her influence is highly dangerous to
sailors. In the only ballad where she is mentioned,
several of the attributes of the Nereids are assigned to her,
such, for instance, as a fondness for the silver poplar—a
trait which is also mentioned in the Popular Tales.[37] In
Elis she is supposed to inhabit a lofty mountain, which
is usually covered with clouds. The character of a storm-

---

[37] Passow, No. 524; Hahn, ii. p. 190.

goddess, in which she thus appears, is thought to have belonged to her also in ancient times, from the fact that she is regarded as the mother of Scylla, who is the personification of the whirlpool.

Besides the Lamia, other frightful conceptions of the ancient Greeks maintain their hold on the popular imagination. The Empusa is still heard of in the upper valley of the Spercheius. The belief in Mormo and Gorgo is also occasionally found; and Gello, another child-eating monster of classical times, continues to be feared by anxious parents, and children that go into a decline are said to be "eaten by Gello."[38] But the most remarkable of all these superstitions are those which relate to the plague. This scourge is conceived of under two different figures. One of these, the outline of which appears to be ancient, though the details are modern, represents it as a blind woman, who passes through the towns from house to house, destroying whomsoever she touches; but, as she can only move through the rooms by feeling her way along the walls, those who are careful to keep in the middle of the apartments are safe from her approaches. In the other, which is still more widely spread, it appears as three women, who make their way through the cities in company, to sweep off the inhabitants. One of them carries a large roll of paper, the second a pair of sharp scissors, and the third a broom. They enter together into the houses of their victims, and the first inscribes their names in her book, the second wounds them with the scissors, and the third sweeps them away.[39] In this last idea we cannot fail to recognise a perverted image of the three Fates. The other superstitions which exist

---

[38] See refs. in Wachsmuth, pp. 57, 77. From the same authority (pp. 30, 31, 55-57) many of the facts relating to the Lamia are derived.

[39] Fauriel, Pref. p. lxxxiii.

concerning those remarkable figures of ancient mythology, especially those which relate to marriage, have been already noticed in connection with Mount Olympus, round which they centre.

The Furies do not appear among the conceptions of modern times; but a custom still remains which bears witness to the former belief in them. When I was first travelling in Greece, I was surprised by seeing a man, who had worked himself up into a passion while talking to me, stoop down and strike the ground violently with his hand. This gesticulation is not unfrequently used by the Greeks when greatly excited, and is in itself very impressive; but, though no further idea is associated with it in the minds of the people, yet in old times it was intended to summon up the Furies from below for purposes of vengeance. Thus, when Althæa, in the 'Iliad,' curses her son Meleager, we are told that she—

> —— "prayed to Heav'n above, and with her hand,
> Beating the solid earth, the nether powers,
> Pluto and awful Proserpine, implored ——" [40]

and in answer to her call "the Erinnys heard from Erebus." The same tendency to euphemism which caused the Greeks to call these goddesses by the name of Eumenides, or "the Merciful," also survives at the present day. We have noticed that the Nereids are called for this reason "the good ladies;" besides this, the small-pox,—which, like the plague, is probably personified, as it is by the Hindoos,[41]—is called Eulogia, or "Blessing," and the ill-omened owl bears the name of Charopuli, or "bird of joy." [42]

---

[40] 'Iliad,' ix. 568 *supp*. (Lord Derby's translation). Compare Wachsmuth, p. 64.   [41] Col. Sleeman, in 'Grote's Greece,' i. p. 299, *note*.
[42] Fauriel, Pref. p. lxxxiv.   Wachsmuth, p. 106.

We now come to an important part of our subject—the ideas of the Modern Greeks respecting death and the state of the dead. And here at the outset we may say broadly, that their beliefs in this respect, as far as we have the means of judging of them, are absolutely and entirely pagan. In the numerous ballads which relate to it there is not a trace of any features derived from Christian sources, while the old classical conceptions are everywhere manifest. It may be said, indeed, that in any country the views on the subject of religion which might be gathered from a collection of popular songs, would be of a very questionable description, and would not fairly represent the beliefs of the people; but this objection does not apply to the Modern Greek ballads, as they are the simple and straightforward expression of the ideas of an unlettered people on the points to which they refer. In this collection some of the songs are intended for Christian festivals, others are dirges to be sung at funerals, and others relate to subjects akin to these, but in none of them does the belief in a resurrection or a future judgment make itself apparent, and there is only one passage, as far as I know, which implies the final punishment of sin. That the people at large have no knowledge of those doctrines, it is hard to believe; but at all events they have not a sufficiently firm hold on their minds to come prominently forward, and they certainly have not succeeded in expelling the old heathen notions. And if most of the figures which we associate with the Inferno of the Greeks, such as Pluto, Proserpine, Hermes, Cerberus, &c., are now wanting, it should be remembered that in ancient times the popular conceptions of such a subject were, in all probability, much simpler than the elaborate scheme which is found in the poets.

Hades, Tartarus (τὰ τάρταρα τῆς γῆς), and "the lower world" (ὁ κάτω κόσμος), are the names by which the abode of the departed is commonly known. It is a dank, dismal region beneath the earth, with a flight of steps leading down to it from the upper air, and closed by a well-guarded door above; it is destitute of light and water, a "chilling region of thick-ribbed ice" (κρυοπαγωμένο), the ideal of all that is unlovely in the eyes of an inhabitant of a southern clime. There the spirits of the departed—men and women, young and old, good and bad—are all enclosed; the women sing dirges, the young men shed tears;—"were we not young?" they exclaim, "were we not Palicars? did we not make voyages in vessels of Psara?" All alike pine for the bright sunshine of the upper world, and for the joys they have lost: the young mother longs to revisit the infant she has left behind her, and which, she fears, must be crying for her breast; the maiden yearns to return to her mother and her brothers, who are sorrowing for her sake. Many are the vain attempts they make to escape from that dreary dungeon. Nowhere is death represented as a rest after toil, or as coming with a friendly aspect to relieve the sorrowful: it is all the old Greek idea, so constantly repeated in the epitaphs of the Anthology, of the loss of all things worth having, the entrance to the dismal world. "My daughter," exclaim the parents of a dying girl, "in Hades, whither thou hastenest to descend, no cock crows, no hen clucks cheerily; there no water is found, there no grass grows. When thou art hungry thou can'st not eat; when thou art thirsty thou can'st not drink; when thou longest for rest thou can'st not take thy fill of sleep. Abide, my daughter, in thy home! abide in thy parents' house!" "My father!" she replies, "my beloved mother! it is of no avail: yesterday I was married, yesterday at late eventide: Hades is my husband, the tomb adopts me as her daughter."[43] How closely the feeling of hopelessness in death, which is expressed in this and numberless other passages, corresponds to the saying of Achilles in the 'Odyssey,' where Ulysses meets him in the world of shadows—

> "'Scoff not at death,' he answered, 'noble chief!
> Rather would I in the sun's warmth divine
> Serve a poor churl who drags his days in grief,
> Than the whole lordship of the dead were mine.'"[44]

The central figure in this lower-world mythology is the terrible Charon (Χάρος, Χάροντας). He still wears the same aspect which he wore of old, that of a surly fierce-eyed old man; but his office is changed, for he is no longer the ferryman of the dead, "the pilot of the livid lake," but rather discharges the functions which were formerly assigned to Hermes, of leading the souls down to the lower world. Mounted on a spectre horse, which here, as in other mythologies, symbolizes the long journey into the silent land, he is seen sweeping over the mountains, which darken at his approach, accompanied by the train of spirits—the young in front of him, the old behind him, the children hanging from his saddle. He thus corresponds in many ways to our conception of Death, only he is far more vividly realised. In fact, no other idea presents itself so forcibly before the minds of the people, or has so powerful an influence over them. "May Charon take you," is a form of curse among the Greeks. An old shepherd, who had been reciting one of the ballads on this subject to Ulrichs, on the side of Parnassus, asked him earnestly whether he thought Charon

---

[43] Passow, No. 374. See also the 'Myrologia,' and 'Carmina Charonea' in that collection generally, and especially Nos. 359, 369, 371, 421.

[44] Homer's 'Odyssey,' xi. 488-91 (Worsley's translation).

had ever let go one whom he had got into his power.[45] His appearance is described in a manner calculated to inspire the utmost awe—"his countenance like lightning, his shoulders like two mountains, his head like a castle."[46] When he starts on his hunting, he shoes his steed, that there may be no delay on the road. His victims he either shoots with an arrow, or bites in the form of a serpent, or swoops down upon as a bird of dark plumage. Those who make light of his power are certain to be visited by his indignation. Young girls, who in the pride of their beauty venture to despise him, immediately become his prey. He presents himself before young men in their full vigour, and announces to them his errand, and when they refuse to give up their lives without a struggle, he wrestles with them in a threshing-floor; the contest is often long and fierce, but as surely as evening comes, the young man's strength begins to fail, and Charon seizes him by the hair, and drags him away captive. More impressive, because more ideal, than these descriptions of the ballads, is a story which is related in one of the Popular Tales, and which is remarkable for the mixture of pagan with Christian ideas which it contains—if Christian they can be called. In this, a woman, who offers a large quantity of incense on a mountain-top, attracts the attention of the Lord Jesus, who sends his angel from heaven to inquire what recompense she desires for so good a deed. She asks for the power of seeing what happens to a man's soul at his death. Her request is granted, and a few days after, when a young man dies in the neighbourhood, she goes to the spot, and sees Charon struggling to take his soul from him by force, while the other resists with all his might. This secret she after-

---

[45] Ulrichs' 'Reisen in Griechenland,' i. p. 133.   [46] Passow, No. 428.

wards betrays to her husband, and has to pay the penalty of her loquacity by her death.[47]

Charon is also the gaoler of the dead in Hades, and in this character his merciless nature is no less seen. The bolder spirits conspire to make their exit; they even steal his master-key: but "hush!" they exclaim, "Charon will hear us!" and Charon does hear them, and never fails to prevent their escape. "Charon is a clever knave, the first of Clefts; he knows the stratagems of Clefts, and the wiles of women." The following ballad describes the pathetic appeal of a maiden in Hades to three daring men who were planning their escape:[48]—

"How happy are the bonny hills! how happy are the meadows!
They suffer not from Charon's might, nor fear th' approach of Charon,
But feed the flocks in summer-time, and bear the snows in winter.
Three valiant heroes took in hand to force their way from Hades;
The first would go in gladsome May, the second in the summer,
The third in autumn would escape, when grapes are in the vineyards.
A fair-haired maiden spake to them, there in the world of shadows:
'Take me too with you, valiant men, up to the air of daylight.'
'Thy garments rustle, maiden fair, thy flowing tresses flutter,
Thy slippers make a flapping sound, and Charon will perceive us.'
'O no—my dress, I'll put it off; my tresses, I will cut them;
And at the bottom of the stairs I'll leave my flapping slippers.
O take me with you, valiant sirs! take me too to the daylight,
That I may see my mother dear, how for my sake she's grieving;
That I may see my brothers too, how for my sake they're weeping.'
'Thy brothers, maiden—in the dance, lo! they are gaily dancing!
Thy mother, maiden—in the street, lo! she is gaily talking!'"

Here is the irony of circumstances indeed! In another ballad a messenger-bird is described as issuing from the gloomy dungeon, with crimson talons and dusky feathers —his talons crimson with blood, his feathers dusky with the shadow of the grave; mothers, and sisters, and wives

---

[47] Hahn, No. 60.   [48] Passow, No. 420.

crowd round him, bearing offerings, and enquiring for news of their loved ones; but he has no story to tell, save of the conquests of Charon, and of the spoil he brings from the upper world.[49] The tent in which he dwells is red without and black within; its pegs are the hands of heroes, its ropes the plaited tresses of women's hair.[50]

If we examine the ancient belief about Charon somewhat more closely, we shall perhaps find reason to conclude that it was not altogether so far removed from the modern conception. At what time it was introduced into Greece we do not know for certain, but the name does not occur in Homer, and at that early period there was no idea of a ferryman carrying the souls across an infernal river. Otfried Müller suggested that the Charon of the early Greek traditions may have been a great infernal deity,[51] and in the late poems of the 'Anthology' he certainly appears as carrying off the dead from the earth.[52] The origin of the belief is to be referred to Egypt. Thus Sir Gardner Wilkinson says: "Of Charon it may be observed that both his name and character are taken from Horus, who had the peculiar office of steersman in the sacred boats of Egypt; and the piece of money given him for ferrying the dead across the Styx

---

[49] Passow, No. 410.   [50] Ibid., No. 433.
[51] 'Etrusker,' ii. p. 100. He remarks that the Charonian staircase of the Greek theatre, by which the shades of the dead entered from below, implies a conception of wider import than that which we usually find; and also that the Latin *Orcinus* was translated by Χαρωνίτης.
[52] 'Anthol. Gr.,' vii. 603 :—

     Ἄγριός ἐστι Χάρων. πλέον ἤπιος. ἥρπασεν ἤδη
     τὸν νέον. ἀλλὰ νόῳ τοῖς πολιοῖσιν ἴσον.

Ibid., 671 :—

     Πάντα Χάρων ἄπληστε, τί τὸν νέον ἥρπασας αὔτως
     Ἄτταλον; οὐ σὸς ἔην, κἂν θάνε γηραλέος;

appears to have been borrowed from the gold or silver plate put into the mouth of the dead by the Egyptians."[53] And again, "One of the principal duties of Horus was that of introducing the souls of the dead into the presence of Osiris, after they had passed the ordeal of their final judgment. He also assisted Anubis in weighing and ascertaining their good conduct during life, previous to their admission into the august presence of his father."[54] The corresponding belief in Etruria seems to have been derived from the same source, and attributes to Charon a more widely extended office than is found in ordinary Hellenic mythology. Mr. Dennis, in his work on Etruria, gives the following description of that divinity, as he is depicted on the Etruscan vases and monuments:—
"Like the ferryman of the Styx, the Etruscan Charun is generally represented as a squalid and hideous old man, with flaming eyes and savage aspect; but he has, moreover, the ears and often the tusks of a brute, and has sometimes negro features and complexion, and frequently wings—in short, he answers well, cloven feet excepted, to the modern conception of the devil. He is principally distinguished, however, by his attributes, chief of which is the hammer or mallet; but he has sometimes a sword in addition, or in place of it, or else a rudder, or oar, which indicates his analogy to the Charon of the Greeks; or a forked stick, perhaps equivalent to the *caduceus* of Mercury, to whom as an infernal deity he also corresponds; or, it may be, a torch, or snakes, the usual attributes of a Fury." The writer then goes on to tell us that he is most frequently introduced as intervening in cases of violent death; that he is often represented as the messenger of Death, leading or driving the horse on

---

[53] Wilkinson's 'Ancient Egyptians,' v. p. 433.   [54] Ibid., iv. p. 401.

which the soul is mounted; or accompanying the car on which the soul is seated; or attending the procession of souls on foot into the other world; lastly, that he is the tormenter of guilty souls, and that his hammer or sword is the instrument of torture.[55] The close resemblance which the Etruscan Charun bears to the modern Greek conception confirms the idea that the latter has descended from ancient times.

In connection with Charon and the state of the dead there are two customs which deserve to be mentioned as being of classical origin—that of putting a piece of money (ναῦλον, δανάκη) into the mouth or hand of the corpse at burial; and that of feeding the *manes*. As to the former of these Mr. Newton says—" The present Archbishop of Mytilene, who has been much in Macedonia, told me that in that uncivilised and remote part of the Turkish empire the Greek peasants still retain the custom of putting a ναῦλον in the mouth of the dead. Wishing to put an end to this relic of Paganism, he explained to them that the coin they used for the purpose being a Turkish para, and being inscribed with a quotation from the Koran, was consequently quite unfit to be placed in a Christian tomb."[56] From Von Hahn and others we learn that the same practice exists in many parts of Greece, in Asia Minor, and in Albania, where " before the corpse leaves the house where the person died, a para or some other coin is placed in its mouth, unless it happens to wear a silver ring."[57] We have already noticed it among the Wallachians, who probably inherited it from the Romans; but there it is placed

---

[55] Dennis's 'Cities and Cemeteries of Etruria,' ii. pp. 206, 207.
[56] Newton, i. p. 289.
[57] 'Albanesische Studien,' i. p. 151, and p. 199, *note*.

in the hands of the corpse. In Asia Minor it is even called "passage money" (περατίκιον).[58]

As to feeding the *manes*, the custom still continues of placing food on the grave at stated periods after death, and pouring libations of wine; but the former part of the observance has now in most places been superseded by a distribution of the offerings among friends and neighbours. More striking is the story which was related to Mr. Newton, that "one day when the body of a young girl was lying in a church waiting to be interred, the Archbishop observed a woman slipping a quince into the bosom of the corpse. On questioning her, she confessed that she had secreted this offering in the hopes that the dead girl might convey it to her own son, who had died about three weeks before. The Archbishop was greatly scandalized; and telling the poor woman that such superstitious practices might cause her own death, gave the quince to a child, who ate it with happy unconsciousness that he was robbing the *manes* of their due."[59]

I might proceed to give an account of numerous other customs which have descended to the modern Greeks from antiquity; but this would be foreign to my present purpose, and would weary the reader by a multiplicity of details. A few of them, however, may be simply enumerated. Throwing back the head (ἀνανεύειν) is still the regular way of giving a negative answer throughout the Levant. The *astragalus* is used to play with very much as it was in ancient times. Throwing an apple is used as a sign to express love, or to make an offer of marriage. Sugar-plums, nuts, and similar gifts, are showered from windows by friends over the bride and

---

[58] Wachsmuth, p. 118.   [59] Newton, ii. p. 11.

bridegroom in marriage processions. The capture of the bride still exists here and there, as it did in ancient Sparta. Great importance is attached in certain ceremonies, especially those connected with marriage, to their being performed by young men both of whose parents are living. The custom of singing songs to welcome the returning swallows in spring is also maintained, and the ballads which are used on this occasion wear a close resemblance to one of the same kind which has come down to us from the classical period. Similarly it has been shown that the seclusion in which Greek women are kept is derived, not from the example and influence of the Turks, but from the practice of ancient times. For some of these, and for numerous other customs, together with the prototypes from which they have descended, the reader is referred to the notes and appendices to Wachsmuth's excellent essay on the subject, which has already been frequently noticed.

But the most interesting question connected with this point is the way in which Pagan practices have been adapted to, or embodied with, the beliefs and ceremonies of the Christian church. That this took place in the Eastern communion as well as in the West, and in our own northern clime, does not admit of a doubt. Several of the ancient divinities were replaced in the popular imagination by Christian saints, to whom corresponding powers and attributes were attached, and Christian festivals succeeded in supplanting those of the heathen gods, in no slight degree, by means of the features in which they were made to resemble them. In particular, the numerous points of correspondence between the observances of Good Friday and Easter Day in the Greek church, and those of the Eleusinian mysteries, have attracted especial notice. How natural such a change was, and how little it was associated with ideas either of trickery or of irreverence by the early Christians, may be seen from De Rossi's great work, 'Roma sotterranea Christiana,' in which it is shown to how great an extent Pagan fables and Pagan symbols were employed in the Catacombs as a means of representing Christian truth. "It may be a kind of shock," a recent writer remarks, when speaking of this subject,[60] "to find what unceremonious and unflinching use was made even of Pagan emblems in aid of Christian symbolism; but there is no cause for such alarm. Men's minds were preoccupied by the old myths, and their new-found faith and hope could not at once sponge out the recollections of their childhood, however it might rule their wills." "Whether Christian artists looked on the tales of Deucalion and Hercules as foreshadowings of the truth in heathen minds, or not, they made use of them. The pictures of Noah in the Catacombs present an exact analogy with medals of Septimius Severus stamped with the deluge of Deucalion. The history of Jonah is perhaps the most frequently-chosen subject of all. No doubt our Lord's reference to it, as a type of His own resurrection, had much to do with this. But the history and its representations are strongly connected with those of Hercules, Jason, Hesione, and Andromeda. The last fable, in particular, had for its scene the coast or the city of Joppa, and was thus on common ground with Jonah's history. And it seems to have caused no painful feeling of irreverence in the minds of secret worshippers prepared to die the martyr's death, to recognise, in the pictures of the deeds of saints and prophets, adaptations of the ideals of well-taught Paganism." When this spirit

---

[60] Mr. Tyrwhitt, in the 'Contemporary Review,' No. i. pp. 73, 74.

prevailed during the first struggle between Christianity and the heathen world, we can well understand how, at a later period, when the new religion had become dominant, Pagan festivals would be adapted, and numerous Pagan customs retained even in Greece, notwithstanding the destruction of temples and statues. The whole subject is one that deserves a fuller investigation than it has yet received. When the materials by which it may be elucidated have been sufficiently collected and critically examined, a new chapter will be added to ecclesiastical history.

APPENDICES.

(A.)

ON THE TOPOGRAPHY OF TROY.

(B.)

GREEK INSCRIPTION FOUND AT MONASTIR.

(C.)

ON THE RETREAT OF BRASIDAS FROM LYNCESTIS.

(D.)

ON THE EGNATIAN WAY.

(E.)

THE BIRTHPLACE OF JUSTINIAN.

(F.)

ON THE MARCH OF A ROMAN CONSUL ACROSS MOUNT OLYMPUS

(G.)

ON THE SITE OF DODONA.

APPENDIX A.

ON THE TOPOGRAPHY OF TROY.

The difficulties which beset the question of the position of ancient Troy and the neighbouring localities do not arise from insufficient knowledge of the district, or from want of careful investigation of the subject. Few classical sites have more attracted the attention of intelligent travellers; and though something still remains to be done in the way of excavation, yet, in respect of the features of the ground, the English Admiralty Survey has left nothing to be desired. The passages from ancient authors which bear on the subject have been brought together, and carefully sifted and compared. The traditions have been collected and their historical value critically examined, especially by that accomplished critic Welcker. The modern authorities—whose writings amount to a respectable literature—have been enumerated by the same author and by Mr. Maclaren. All the sites for which any claim can be put in have found their advocates: and for and against these respectively almost every available argument has been employed. Yet, after all this, none of the views that have been advanced are free from considerable difficulties; and we cannot help feeling that their supporters have somewhat ignored the force of these, and underrated the views of their opponents; though this defect has its corresponding advantage, in enabling us to see more clearly what may be said for the different sides. But, as I have already observed,[1] the Homeric topography does not appear to be imaginary; nor is there any reason to believe that the ground has so far changed as to render the search hopeless. And when we consider that a question so comparatively simple as that of the position of the harbours of Athens has only recently been decided, there is no cause to despair of arriving ultimately at a satisfactory solution of the present investigation.

The way may, to some extent, be cleared by putting aside those views which may now be regarded as impossible. Such is the supposition of Belon, that Alexandria Troas was the site of ancient Troy.

---

[1] *Supra*, vol. i. pp. 25 foll.

Such, again, is that of Wood, the first traveller who bestowed much attention on the subject, who, finding that the Scamander was said to rise in front of the city, and that the river which best corresponded with it had its source in the upper part of the chain of Ida, violated all the probabilities of the case by placing Ilium in that remote region. Nor is much more to be said for Rennell's view, though it is supported by greater learning and more careful observation; since, relying on a passage which describes Ilium as having been built in the plain,[2] he would place the site of the city in the level ground, notwithstanding the almost universal custom of the inhabitants of these countries to build in elevated positions, and in defiance of the numerous epithets which imply that Troy was in a steep and lofty situation. The places whose claims remain to be discussed, after these have been put out of the question, are the Bali-dagh, or hill behind Bunarbashi; Atchi-keui, the site of Pagus Iliensium; and Hissarlik, where are found the remains of Ilium Novum.[3]

A further step in advance will have been gained, if we can persuade ourselves that the ancient traditions on the subject are to be wholly disregarded: this is Welcker's view,[4] and he seems to be justified in his conclusion. The period that intervened between the composition of the Homeric poems and the historic age was a long one—quite long enough to admit of the position of the city being forgotten; for it is surprising how soon sites of great and permanent interest become matters of doubt, as we see in the case of the Holy Sepulchre. For some time, too, the country was inhabited by a barbarous Thracian tribe, the Treres,[5] who would not be likely to care for the former history of the place. Besides this, the traditions which existed at a later time with regard to the site were themselves conflicting; so that whilst most authorities place it at Novum Ilium, some—as, for instance, Strabo, who follows Demetrius of Scepsis, a native of the country, who lived about 180 B.C.—prefer the claims of Pagus Iliensium.[6] And if anything more were wanted to invalidate the traditions

---

[2] 'Iliad,' xx. 216 :—

οὔπω Ἴλιος ἰρὴ
ἐν πεδίῳ πεπόλιστο.

(See Rennell, 'Observations on the Topography of Troy,' p. 115.) An examination of the passage will show that what is intended by ἐν πεδίῳ is 'in the lower country,' or 'towards the plain,' in contrast to the higher regions, where the inhabitants originally dwelt.

[3] See map on p. 23 of vol. i.

[4] 'Kleine Schriften,' vol. ii. 'Ueber die Lage des Homerischen Ilion,' pp. iv-xiii.   [5] Strabo, xiii. 1, 8.   [6] Ibid., xiii. 1, 35.

in favour of any particular spot, it would be found in distinct statements which occur in some ancient writers, to the effect that the old city was utterly destroyed and deserted.[7] This point is of great importance: because the supposed evidence of tradition forms the strongest argument in favour of Novum Ilium, and has been put forward as such by an able advocate of that place, Dr. von Eckenbrecher.[8] The people of Ilium not only pointed out in the neighbourhood of their city all the features, even the most minute, which are mentioned in the 'Iliad,' but also maintained that Troy had never been destroyed, but had continued to exist in the same place all along. This is nothing more than what we might expect under the circumstances. When the new town received the name of Ilium, it may have been given with the hope, or to suggest the idea, that it might some day rival the ancient city; but, whatever was the origin of the appellation, it would almost certainly come to pass, if the place rose to eminence, that it would claim a direct connexion with Priam's capital, and discover a correspondence between the objects in its neighbourhood and those which Homer describes.[9] After this belief was once well established, there would be no difficulty in securing its general reception. Powerful conquerors, like Xerxes and Alexander, on visiting the spot, would naturally accept the prevailing tradition, and would not go out of their way to offend the citizens by questioning the genuineness of their pedigree. And in like manner the majority of the Greeks, and still more the Romans, who were eminently uncritical in literary and topographical research, would have no reason to question so harmless a claim. But if there is reason to believe that the story of the connexion of the two places had a fictitious origin, any amount of consent in favour of it among ancient writers is valueless as evidence.

The position of the three places whose claims remain to be discussed has been already described in my chapter on the city and plain of Troy; and to the same part of my narrative I will refer my readers

---

[7] Lycurgus the orator (quoted by Strabo, xiii. i. 41), addressing the Athenians, says: τίς οὐκ ἀκήκοεν, ὡς ἅπαξ ὑπὸ τῶν Ἑλλήνων κατεσκάφθη, ἀοίκητον οὖσαν; similarly Lucan writes ('Pharsal.' ix. 968):—

—'tota teguntur
Pergama dumetis: etiam periere ruinæ.'

See also the passages from Homer, quoted by Strabo, as above.

[8] In the 'Rheinisches Museum' for 1843, pp. 28 foll.

[9] οἱ νῦν Ἰλιεῖς φιλοδοξοῦντες καὶ θέλοντες εἶναι ταύτην τὴν παλαιάν—is Strabo's allegation against them (Strabo, xiii. i. 25).

for a description of the neighbouring district, for the general features of the topography of the 'Iliad,' and for the arguments in favour of the site above Bunarbashi. Let us now proceed to a somewhat more minute examination of some of the objects in the plain.

The first point which naturally attracts our attention is the rivers. Besides the principal streams of the plain—the Mendere, the river of Bunarbashi, the Kimar, and the Dumbrek—there are numerous watercourses, which are for the most part dry in summer, but in winter contribute in no slight degree to the frequent inundations. The beds of these are carefully marked in the Admiralty Survey, but they do not materially affect the topography, though the investigation of them has been of great service in removing misconceptions with regard to the courses of the rivers. The Mendere is now almost universally allowed to be the Scamander. The similarity of the names is sufficiently striking, for the loss of the first syllable is a common form of corruption in names, as in Salonica for Thessalonica; but we may not be too confident about this derivation of the modern name, because two other rivers of Asia Minor, the Mæander and Cayster, are now known as the Greater and Lesser Mendere. However, the point is one of small importance; for if the connexion was shown, it would prove no more than what we know already, viz., that this river was called Scamander in the historical times in Greece; the only question being whether it bore that name in the Homeric period. Great difference of opinion has existed about its course. Forchhammer, who from having accompanied the English Survey in the examination of the district has every right to be heard on this subject, confidently maintains that it has never changed its direction; and the well-defined bed in which it runs, notwithstanding the neighbourhood of other watercourses, makes this highly probable. But the majority of the critics incline to the other view, and believe that in the lower part of the plain in ancient times it trended across again from the western to the eastern side, and flowed into the sea close to the Rhœtean promontory. The arguments, however, in favour of this are not satisfactory. Leake's idea that the ground throughout the plain sloped downwards from east to west, and that consequently the river had made its way across to the western side,[10] has been overthrown by the Survey, which proves that the ground is level from one side to the other. The traces which he believed to have found of earlier channels on the eastern side are in reality the beds of other and separate streams. In the inter-

---

[10] 'Tour in Asia Minor,' p. 292.

pretation of Homer equal difficulties are involved by this supposition. If we place the city of Troy at the head of the plain, the Scamander will have to be crossed twice before the ships are reached; whereas Homer always speaks of one ford only.[11] Again, if it flowed by Rhœteum, there would be no reason for Hector, when wounded at the ships, being carried in a fainting state to the ford, half-way to Troy, to have water poured over him,[12] as the river would be close to the station of Ajax, where he had been fighting. Rennell, who is referred to as the principal authority for this view, states confidently that the Scamander was near the tents of Ajax,[13] but he nowhere proves it; and one of the passages on which he lays the greatest stress admits of a different interpretation from what he puts upon it. Hector, when driving the Greeks from Troy to the ships, is described in one part of his course as fighting "on the left of the whole battle by the banks of the river Scamander;[14] and from this Rennell concludes that the Scamander must have been on the east side of the plain, because the positions of the forces on both sides are described from the point of view of the Greek force and camp, which would place Hector on the right wing of his own army. Even if this view is true (and to me it seems extremely doubtful), the passage proves nothing, because we have no means of deciding whether the events there spoken of are supposed to have happened above or below the ford in the middle of the plain; and if they happened above the ford, the river, on every supposition, would be on the left of the Greeks. Mr. Maclaren endeavours to show, from the state of the soundings at the present day, that the river in Strabo's time must have entered the sea at a point half-way between its present mouth and the Rhœtean promontory, because the deposit of alluvium off that part of the shore projects further into the sea than we should expect; and from this he proceeds to argue that its course lay still further to the east in Homer's time.[15] This argument, however, is based on sand and water—an eminently unsafe foundation. As, besides the ordinary current of the Hellespont which flows from east to west, there is another current along the shore in these parts from west to east, it is very difficult to determine by what exact process the submarine deposits may have been formed; and this difficulty is increased by the presence of another river, the Dumbrek, which has contributed in some measure to the formation of the alluvium. Against all these speculations is to be set the express

---

[11] 'Iliad,' xiv. 433; xxi. 1; xxiv. 692. [12] 'Iliad,' xiv. 433.
[13] 'Observations on the Topography of Troy,' pp. 87 foll.
[14] 'Iliad,' xi. 498. [15] 'The Plain of Troy,' pp. 46 foll.

statement of Strabo, that the Scamander in his time flowed into the sea over against Sigeum.[16] This does not necessarily imply that it kept close to the hills during the latter part of its course; and his statement that it was joined by the Simois (Dumbrek) in front of Novum Ilium, and that its mouth was 20 stades from that city, seem to imply that it did not do so; but it certainly must mean that it was considerably nearer to the western than the eastern side.

We now come to the river of Bunarbashi, which has been already described as a narrow perennial stream of clear water, rising from numerous sources near the village from which it takes its name, and flowing at the present day through a cutting in the hills into the Ægean, though the channel still remains by which in former times it joined the Mendere. Whether this cutting is an ancient or a modern work, it is not easy to decide. Ulrichs states,[17] that he found that both the Greek and the Turkish peasants in the neighbourhood believed it to have been dug by prisoners from the Morea after the conquest of that country. On the other hand, Leake,[18] Forchhammer, and other good judges, pronounce it ancient, especially on the ground of the difficulty of executing it. The last-named author remarks : " If travellers had inspected this work closely, they would have seen that it is cut for a considerable distance through a thick layer of rock, and must have been convinced that the turning a poor Turkish mill could never have offered sufficient inducement to such a laborious undertaking, or repaid the time, outlay, and labour, expended upon it. Besides, the circumstance that a great part of the plain on the west of the Mendere is preserved from constant inundation by this diversion of the water, sufficiently indicates its real object. No Turk or Turkish governor ever undertook so bold and toilsome an operation."[19] It is no answer to this to say, with Mr. Maclaren, " As if the people who built the splendid mosques of Constantinople would have shrunk from the task of cutting a gutter in a soft tertiary limestone; "[20] for the Turks have executed remarkably few works, small or great, out of their chief cities. It is not unlikely that the operation of which the peasants had heard was the cleaning out and restoring of the ancient channel, and

---

[16] Strabo, xiii. i, 34. Similarly the Scholiast on 'Iliad,' ii. 467, says—
ὁ δὲ Σκάμανδρος καταφερόμενος ἀπὸ τῆς Ἴδης μέσον τέμνει τὸ ὑποκείμενον τῇ Ἰλίῳ πεδίον καὶ ἐπὶ τὰ ἀριστερὰ ἐκδίδωσιν εἰς τὴν θάλασσαν.

[17] ' Reisen in Griechenland,' ii. p. 265.

[18] ' Disputed Questions of Ancient Geography,' p. 33.

[19] ' Journal of the Geographical Society,' vol. xii. pp. 37, 38.

[20] ' Plain of Troy,' p. 129.

the story related by one traveller is to this effect.[21] An additional interest has been imparted to the question by a much-disputed passage of Pliny relating to the Troad. When enumerating the places along the west coast of the Troad, from south to north, he mentions " the town of Nea, the Scamander, a navigable river, and the promontory on which once stood the town of Sigeum. Next comes the Harbour of the Achæans, into which the Xanthus flows, united with the Simois, after first forming the marsh called Old Scamander."[22] The impression which this passage leaves on our minds certainly is, that Pliny believed that the Scamander flowed into the sea south of Sigeum, just where the river of Bunarbashi now flows; that another river, the Xanthus, which can be none other than the Mendere, joined by the Simois—which at that time was the name of the Dumbrek—made its way through the plain into the Hellespont; and that above the junction of the two a communication had once existed between the Xanthus and what he calls the Scamander, which was now converted into a marsh. Still, even in an inaccurate author like Pliny, it would be surprising to find a stream which can only just float a small boat described as a " navigable river;" and it is certain that at that time the Mendere was generally called Scamander. The cutting may, perhaps, have then existed; but for the Homeric topography the question is of slight importance, and all that can be deduced from this passage is, that in Pliny's time there may possibly have been conflicting traditions as to the names of the rivers. But whatever may have been the antiquity of the work, of its usefulness there can be no doubt; nor was it the only thing of the kind in this district, for a similar cutting exists nearer Yenishehr, though now partially filled up, and was evidently intended to relieve the lower part of the plain from water in time of inundation. When seen from the Ægean this forms a striking object, as the hills appear to be cut through to their base.

One writer, indeed—Mr. Maclaren—goes somewhat further than others in this matter, and maintains that the bed of the Bunarbashi river is artificial from first to last, and did not exist at all in ancient times; but that the water which issued from the springs converted the whole of the upper part of the plain on the left bank of the Mendere into a marsh.[23] There is an antecedent improbability in this supposition, for the unhealthiness and unproductiveness of marshes at once suggest the necessity of drainage; and thus we find that the channels

---

[21] Barker Webb, 'Topographie de la Troade,' quoted by Welcker, 'Kleine Schriften,' ii. p. xlvii.   [22] Pliny, i. 33.
[23] 'The Plain of Troy,' pp. 151, 152.

for carrying off the waters of the Copaic and Alban lakes were amongst the earliest works executed by the Greeks and Romans. If an ancient city existed near the springs, it is almost certain that they would have been prevented from overflowing the neighbouring country. Besides this, though the upper part of the banks of the Bunarbashi river is evidently artificial through a great part of its course, yet there is no reason for supposing that its deep bed is so, and the direction which its waters take at the foot of the hills is perfectly natural, as they flow that way from the very first, as soon as they issue from the easternmost group of fountains close to the village of Bunarbashi.

The identification of this stream with the Scamander is an instance of what we have already noticed, the tendency of topographers towards adapting their whole scheme to one single passage. When Lechevalier first discovered the springs, and observed how well they corresponded with those which Homer has spoken of as sources of the Scamander,[24] he at once concluded that they must be the head-waters of that river, and that all the characteristics attributed to it must be applicable to the stream they formed. Colonel Leake supported the same view with much ingenuity, laying great stress on the value attached to a perennial supply of water in ancient times, and thus explaining the deification of this small river, and its being regarded as the most important stream in the district.[25] Now it is true that the estimation in which a river was held did not depend on its size, as we see in the case of the Ilyssus and Cephisus in the plain of Athens, and other famous streams of antiquity; but in the present instance this argument does not apply. We may claim, indeed, on this ground, a place of honour for the river of Bunarbashi, but not the first place, for that must certainly be conceded to the Mendere, on account of its position in the plain, its volume of water, and the breadth of its channel. And as to the perennial supply, neither of the two rivers can be said to have much advantage over the other: for though the Mendere is known occasionally to fail, yet the evidence of travellers does not go further than to show that its bed is dry for a month or two once in thirty, or at the most in twenty years; and a Greek who lived in the neighbourhood assured me that, though the water of both is good, they prefer that of the Mendere for drinking. Nor, again, can anything be determined from the account of the conflict between Achilles and the Scamander,[26] where an elm-tree which falls from the banks into the

---

[24] 'Iliad,' xxii. 147.   [25] 'Tour in Asia Minor,' pp. 291, 2.
[26] 'Iliad,' xxi. 245. The passage is referred to by Lechevalier in his 'Description of the Plain of Troy,' Dalzel's translation, p. 85.

stream is described as "bridging" it (γεφύρωσεν), for we are immediately afterwards told that it fell completely within it (εἴσω πᾶσ' ἐριπούσα); and the idea that it spanned it is wholly at variance with the general tenor of the passage, as the description is that of a mighty current, sweeping away heaps of slain in its course. Other points, in which the Homeric description of the Scamander is more applicable to the Mendere than to the Bunarbashi river, have been already enumerated.[27]

If, then, this stream was not the Scamander, and yet was of sufficient importance to deserve notice in ancient times, what name did it bear? The answer is, that if we place Troy on the Bali-dagh above Bunarbashi, it must certainly have been the Simois, as that river is described as meeting the Scamander in front of the city.[28] And in other respects it corresponds to what is said of that river, for it flows parallel to the Scamander for some distance, allowing room for the conflicts which are mentioned as taking place between the two.[29] The objection that in the conflict of Achilles and the Scamander, a quiet stream like that of Bunarbashi would not naturally be called on to assist his brother by bringing down logs and stones[30] appears somewhat hypercritical. The whole scene is conceived in a spirit of poetical exaggeration.

The third river of the plain is the Dumbrek. This flows from the east through a gently-sloping valley, almost parallel to the Hellespont, until it reaches a plain which extends below Hissarlik to the Rhœtean promontory, and opens out into the greater Trojan plain. At one period it joined the Mendere, for Strabo tells us that the two rivers flowed together in front of Ilium;[31] that is, we may suppose, on the other side of the plain opposite New Ilium. Now, however, it takes a different course, and bends round the Rhœtean promontory, entering the sea by an estuary at the extreme north-east angle of the plain; yet a branch of it, which runs off from the main stream near the village of Kum-keui, after being joined by another watercourse—the Kalifatli-Asmak—almost touches the Mendere, towards which it crosses the plain in a diagonal line, and flows into the Hellespont close to its mouth. The date of the separation of the two streams and the alteration of the channel of the Dumbrek may be approximately ascertained, for Ptolemy, whose date is about 150 years later than Strabo, represents them as having separate mouths;[32] and as Pliny, in the passage already quoted, speaks of them as uniting their waters, we may con-

---

[27] See above, vol. i. p. 31.　[28] 'Iliad,' v. 773, 774.　[29] Ibid., vi. 2-4.
[30] Ibid., xxi. 313.　[31] xiii. i. 34.　[32] 'Geog.' v. 2.

clude that the change took place between his time and Ptolemy's, that is, between A.D. 60 and 150.[33] In Strabo's time this stream was certainly called the Simois; but whether this was its original name, or whether it was given by the Ilians, in order to reconcile the position of their city with the Homeric topography, we have no means of determining. The similarity of the names Dumbrek and Thymbrius, or Δύμβραιος,[34] as one of the ancient rivers of the plain was called, is not a little striking; and Forchhammer remarks that "the Turks, who can account for the name of every village in the plain, cannot explain the meaning of Dumbrek, or Dumbrek-keui.[35] It is possible that the original name may have been retained by the rural population, who, as often happens, were not affected by the antiquarian fancies of the inhabitants of the neighbouring city. We should observe, however, that the Thymbrius is not noticed by Homer; and the position of the place Thymbra, which is mentioned in the story of the Dolonea,[36] and which we should expect to be near the river, has been much disputed. To me, however, the most natural interpretation of the passage seems to place it near the Dumbrek valley. The Trojan force at that time is stationed at the "rising of the plain" ($\theta\rho\omega\sigma\mu\grave{o}s\ \pi\epsilon\delta\acute{\iota}o\iota o$),[37] which was not far from the monument of Ilus;[38] and this we know was towards the middle of the plain.[39] The spy Dolon, when captured by Ulysses and Diomede, informs them that part of the allied Trojan host was stationed towards the sea ($\pi\rho\grave{o}s\ \dot{a}\lambda\acute{o}s$), and part towards Thymbra ($\pi\rho\grave{o}s\ \Theta\acute{v}\mu\beta\rho\eta s$), while the Thracians under Rhesus, who had just arrived, were on the extremity of the line separate from the others. The sea and Thymbra appear to be spoken of as points between which the whole force lay encamped, though it is not implied that they touched either of them; in fact, this is impossible in the case of "the sea;" for, as they were in the middle of the plain, this must mean the Ægean, which they could not approach because of the intervening hills; and therefore we may suppose that Thymbra also is only named as a well-known point, in the direction of which the right wing extended. The Trojans and their allies would thus reach across the plain; and though we should expect to find the eastern extremity of their line somewhat south of the Dumbrek valley, yet in a general description the expression "in the direction of Dumbrek" would hardly appear unsuitable. The Thracians then would be supposed to

---

[33] See Maclaren. p. 67.   [34] On this form of the name see Welcker, p. xlii.
[35] 'Geog. Soc. Journal,' vol. xii. p. 44.   [36] 'Iliad,' x. 428-435.
[37] Ibid., x. 160.   [38] Ibid., x. 415.   [39] Ibid., xi. 166, 7.

have marched down the valley of the Dumbrek and joined their friends on the field of battle.

About the fourth river, the Kimar, little need be said, as its position in an angle at the head of the plain removed it from the scene of the conflicts of the 'Iliad.' It seems to correspond to the river which Strabo called Thymbrius.[40]

From investigating the courses of the rivers we naturally turn to the question whether there was a bay in ancient times into which they discharged their waters. The great majority of the authorities answer this in the affirmative; but there are a few who hold the contrary view and maintain that the coast-line, both in Strabo's and in Homer's time, was in the same position as at the present day. Thus Dr. Forchhammer says: "The elevation of the whole slope of the In Tepe ridge (*i. e.*, the Rhœtean promontory), and the deep steep banks of the arms of the Kalifatli-Asmak, in the lower part of the plain, appear to be incompatible with the assumption that such a bay ever existed. The extensive sea-lakes at the bottom of the plain are also, I think, inconsistent with the hypothesis. They are very deep, and it is inconceivable that the plain should be protruded into the Hellespont without their being first filled up. The current of the Hellespont carries off and deposits the sand which the Mendere brings down on the shoal to the left in leaving the Hellespont."[41] To this passage the editor of the 'Geographical Society's Journal' appends the following note: "That the effect of rivers in altering the character of the coast at their mouths has been frequently exaggerated may be admitted without going so far in the other extreme as Dr. Forchhammer has done here. After reading his account of the violent alternations of moist and dry, heat and cold, to which the Plain of Troy is exposed, it is scarcely possible to conceive that it has remained altogether unaffected by them for the long space of three thousand years. . . . . The whole alluvium of the Mendere is deposited between the mouth of the Kimar and the sea. The frosts and thaws, too, to which the Plain of Troy is subject will break up the surface soil, and expose it more to the action of the waters. . . . . The sea-banks of the lagoons at the lower end of the

---

[40] Strabo, xiii. i. 35. That the plain in which this stream flowed should be called Thymbra in Strabo's time, notwithstanding its unsuitableness to the Homeric account, except on the supposition that the neighbouring Pagus Iliensium was Troy, seems to prove that the localities of the district, including most of the rivers, were renamed at a period subsequent to the composition of the Homeric poems.

[41] 'Geog. Soc. Journal,' vol. xii. p. 43.

Plain of Troy are the same as we find the sea casting up on many coasts at a short distance from the shore: in proportion as the plain was extended into the sea, these banks would be formed further back in the sea; the lake would be pushed out as well as the coast-line." At the mouth of the Dumbrek the land is certainly encroaching on the sea. A gentleman residing in the neighbourhood informed me that once in two or three years he used to go with a fishing-party to the mouth of the Karanlik-liman, as the estuary of that river is called. They placed nets across the mouth along the bar, and on each successive occasion they found themselves obliged to fix the stakes for the net somewhat further out than before.

That the point at Kum-kaleh, and the beach between it and the foot of the hill of Yenishehr, did not exist in early times, is almost conclusively proved by the passage of the 'Odyssey' relating to the tumulus of Achilles on the lower slope of that hill, which is said to have stood "upon a projecting beach on the broad Hellespont."[42] Both Herodotus and Strabo[43] express their belief that the alluvium was increasing; and though their individual opinions on such a subject may not be worth much, yet what they say seems to imply that such an impression existed in ancient times. But the latter of these two authors has left us further data by which to decide the question, as he gives 12 stades as the distance from the city of Novum Ilium to the sea at the nearest point—that is, at the inlet by Rhœteum called the Haven of the Achæans—and this is not more than half the distance which we find at the present day. He also speaks of the mouth of the Scamander as being distant 20 stades from that city; so that we must conceive of the coast-line in his time as being as far south as Yenishehr, and in Homer's time it would probably reach still further inland, perhaps within little more than a mile of Ilium Novum. In the 'Periplus' of Scylax[44] there is a passage which is hard to reconcile with Strabo's measurements, where we are told that Ilium was 25 stades from the sea. Mr. Maclaren endeavours to solve the difficulty by supposing that during the period of Roman domination, as the city was favoured by the conquerors, it extended gradually a long distance—as much as eight or ten stades—into the plain, and that the distances given by Strabo were calculated from the outer wall. In Scylax's time (400 B.C.)[45] the city

---

[42] 'Od.' xxiv. 80-84.
[43] Herod. ii. 10. Strabo, xiii. i. 36.   [44] Cap. 95.
[45] This is the date given by Mr. Maclaren; 340 B.C. is more probably correct. See Müller's Prolegomena to the 'Geographi Graeci Minores,' p. xxxvii.

was small, and would not extend beyond the area of the hill of Hissarlik; and consequently, as 25 stades fairly represents the distance of the present line of coast, we may conclude that but little change has taken place, and yet that Strabo's statements are true.[46] This is very ingenious: but it seems far more probable that an author like the so-called Scylax should have made a mistake of numbers, than that all these suppositions should prove true, and that the probabilities of the case in respect of the formation of alluvium should be violated. Add to this, that it is most improbable that writers like Strabo and Demetrius, when arguing against the claims of the Ilians, should leave it in their power to reply, as they certainly would, that the boundaries of their city were different at an earlier period, and that they could claim an additional mile's distance from the sea in the Homeric age.

The next question is, what was the position of the ships and camp of the Greeks? About this point there are three different opinions, and as at least two of these existed among the ancients, it is evident that here, as elsewhere, we may safely disregard the voice of antiquity. Strabo inclines to fix it at the place called in his time Naustathmon, a beach in the neighbourhood of Sigeum; but he mentions that others were in favour of the Haven of the Achæans.[47] The difficulty of the question is increased by our not knowing exactly what appearance the coast-line then presented. Perhaps we may not be far wrong in supposing it to have been something like what we see at the present day, only considerably further from the Hellespont; for Strabo speaks of the river, or rather the two rivers Scamander and Simois combined, as forming marshes and a "blind mouth," or lagoon ($\tau\epsilon\phi\lambda\grave{o}\nu$ $\sigma\tau\acute{o}\mu\alpha$, $\Sigma\tau o\mu\alpha\lambda\acute{\iota}\mu\nu\eta$) at their mouth,[48] and the Karanlik-liman corresponds to the Haven of the Achæans—in fact, it may be its lineal descendant, though now moved much further towards the north. The third view is, that the station of the camp was somewhere intermediate between the other two positions. This has sometimes been pushed so far as to represent that it occupied the whole space from side to side: but this seems impossible, for at the nearest point where the slopes of the Rhœtean and Sigean hills approach one another the distance between them is a mile and three-quarters, and Homer expressly tells us that the shore, though wide, could not contain all the vessels, and that from want of room they were forced to arrange them in more than one line, and even so they occupied the whole of the "long mouth" of the shore which was

---

[46] Maclaren's 'Plain of Troy,' pp. 26-30, and 64.
[47] xiii. i. 36.   [48] xiii. i. 31, 34.

bounded by the headlands.[49] The headlands here mentioned (ἄκραι) need not mean those of Rhæteum and Sigeum, but may refer to any projections of the beach which from some cause or another served as boundaries to the position; but on the other hand, the expression "long mouth" (στόμα μακρόν) which is applied to the shore does not seem satisfactorily applicable to a station at either corner of the bay. In my own mind I incline towards placing the Greek station somewhere in this intermediate position, between the mouth of the Scamander which flowed towards Sigeum, and the Haven of the Achæans.

To this view there are two objections: first, that the Dumbrek would have to be crossed in front of the camp, and that no river is spoken of by Homer as being in that position; secondly, that the Greeks could not have occupied such a station in winter, on account of the inundations. To the first of these it may perhaps be answered that there was no need to mention this river, as it was unimportant to the action of the poem, and the poet is wont to ignore the rivers when it suits his convenience to do so. The second is more formidable; but we may fairly reply to it by asking, where could any position be found not open to the same objection, and at the same time capable of accommodating so vast a host as the Greek army is supposed to have been, with their ships, tents, market-place, and accommodation for chariots, horses, and stores? The truth is, that in the 'Iliad' nothing but the present time is provided for. The previous years of the war are hardly more than alluded to, and the difficulties arising from the supposition of so long a campaign are not taken into consideration. The case is the same with the time of year. It is a tale of the summer, the only season at which fighting is possible in such a district, and from first to last, except during the conflict of Achilles and the Scamander, the plain is dry and dusty, and the rivers fordable without the slightest difficulty. The poet had no need to trouble himself about the fortunes of his heroes during the previous winters. There is also, on this supposition, a completeness imparted to the scene of the war which we otherwise fail to secure. A long, level battle-field, with the besieged city rising on a hill at one end, and the camp and ships of the besiegers resting on the sea at the other—it is exactly the grand and simple conception which is suited to the treatment of an ancient epic. Nor are there wanting arguments from the poem itself to lead us to the same conclusion. When Agamemnon at the Greek ships is represented as seeing the watch-fires in front of Troy ('Ιλιόθι πρό),[50] we feel that the

---

[49] 'Iliad,' xiv. 33-36.   [50] Ibid., x. 12.

two positions must be in some way opposite one another. And when Neptune is described at a later period as obliterating all traces of the Greek camp, by bringing together the waters of all the rivers of the Troad, and sweeping away the walls by the force of their combined streams;[51] though the whole passage is strongly imbued with the marvellous, it is not unnatural to suppose that the scene of this devastation was one to some extent exposed to inundations.

Before proceeding to consider the claims of the different sites to be the representative of ancient Troy, it may be well to notice a point on which some stress has been laid—the pursuit of Hector by Achilles previous to his death. The poet has generally been understood to mean, when he says

$$\text{ὣς τὼ τρὶς Πριάμοιο πόλιν πέρι δινηθήτην}—[52]$$

that the two heroes actually ran round three times outside the city walls: we shall have to enquire presently whether this really is his meaning. Eckenbrecher adduces this passage as a strong argument in favour of the site at Ilium Novum, and against that at Bunarbashi, because it is possible to get round the former, notwithstanding the steep slope of the hills, so that he himself, he says, had crossed the ridge on horseback without difficulty.[53] Ulrichs, on the other hand, considers that it favours his site at Atchi-keui, as against that at Ilium Novum, because the ground in its neighbourhood is so much more level, and consequently a more natural scene for the event described.[54] Hahn, again, who adopts the site on the Bali-dagh, assures us of the practicability of such a pursuit, even along the rugged southern face of that hill; and even argues that the obstacles to be met with there are implied in the poem, because in the metaphor introduced at that point of a fawn chased by a hound, mention is made of glens and brushwood through which they pass, and it must be regarded as a *lapsus calami* on Homer's part if there was nothing in the reality to which this should correspond.[55] This last argument, however, ignores one conspicuous feature of the Homeric similes, viz., that their details are constantly ornamental, and that all that is required is that the main point should be illustrated. The real question, however, is, not whether such a chase was absolutely possible in this or that locality, but whether the nature of the ground was such as to lead the poet to entertain such an

---

[51] 'Iliad,' xii. 17-33.   [52] Ibid., xxii. 165.
[53] 'Rheinisches Museum' for 1843, p. 41.
[54] 'Reisen,' ii. pp. 286, 289.
[55] 'Ausgrabungen auf der Homerischen Pergamos,' pp. 30, 31.

idea. Now it is possible to ride across the ridge between Chiblak and Hissarlik, and to scramble along the precipices of the Bali-dagh; but it is difficult for any one who has seen the two sites to suppose that two men in full armour, however superior to the present race of mortals, would be represented as running three times round either of them; the appearance of the ground is wholly opposed to such a supposition. And generally, wherever Troy may have stood, the difficulty is, how to conceive of such a pursuit having taken place at all round a large city, containing a numerous population, and situated in a steep and lofty position. As Mr. Senior has well observed, "The impossibility belongs to the story itself, not to any particular site."[56] Accordingly, Aristotle has classed it among impossible things, the introduction of which is justifiable in order to produce greater effect.[57]

It is still worth considering whether the passage really implies that the heroes made the circuit of the city. One point in the story certainly seems in favour of it, namely, that after their first start is described, a debate of the Gods is introduced; this kind of interlude being the usual means by which, in Homer, an interval is implied in the action, during which something is supposed to happen away from the immediate scene. In the present instance, however, the object of this may be merely to give time for the three courses, only one of which is described. Several good critics maintain that περὶ in the line quoted above does not signify "round," but "near," as it does also in other places of the 'Iliad,'[58] so that what is intended to be represented is that they made three circular courses near the city. A careful examination of the passage will show that this is very much what is described. It is not implied in the story that anything took place away from the field of battle. The Trojans on their walls, and the Greeks in their ranks, are looking on. The aim of Hector is to make for the city, that of Achilles to prevent him from reaching it, and yet to restrain his own side from attacking him, lest they should deprive him of his lawful prize. The scene lies between the walls of Troy and the Grecian lines; and, as Colonel Leake says, the poet intended to describe that circular course which a person invariably takes when he runs from another, and finds no shelter or advantageous position for defending himself.[59] Perhaps this was Virgil's interpretation of the passage, when

---

[56] 'Journal kept in Turkey and Greece,' p. 174.   [57] 'Poetics,' ch. 25.
[58] E.g. vi. 327.   περὶ πτόλιν αἰπύ τε τείχος μαρνάμενοι. If the interpretation of περιδινηθήτην given in the text is admitted, it must be read as two words, and there is no reason why it should not.
[59] 'Tour in Asia Minor,' p. 305.

in the closing scene of the 'Æneid,' where he has imitated it, he makes Æneas pursue Turnus in a circle in the plain.[60] It is true that from the position of Laurentum, where it took place, he could not have made them run round that city; and that he believed it possible to run round Troy, for he made Achilles drag the body of Hector round it;[61] but yet the coincidence between the Roman poet and his great prototype on this supposition may fairly be regarded as an additional argument in its favour.

The reasons for placing the site of Troy at Atchi-keui are neither numerous nor strong. In the first place there is the authority of Strabo; but this must not be taken for more than the opinion of a learned man in ancient times. Then there is the tradition of the people of Pagus Iliensium, the town which formerly occupied this position;[62] and to this, like the rest of the traditions of the Troad, we need not attach any importance. Next, Plato's statement is adduced that Troy was built on an inconsiderable hill ($\dot{\epsilon}\pi\grave{\iota}$ λόφον οὐχ ὑψηλόν),[63] which is contrary to the description given in Homer. But the argument on which the greatest stress is laid is drawn from the look-out place of Polites,[64] the spy of the Trojans. As he is sent from Troy to watch the movements of the Greeks, it is argued that the Greek camp could not have been visible from Troy, and it would not be within sight at Atchi-keui, while it would be so at Bunarbashi and at Hissarlik. The tumulus of Æsyetes, therefore, which was his position, is placed at a point on the hills halfway between Atchi-keui and Chiblak. There is no need, however, to lay so much stress on this point; the requirements of the case are provided for if we suppose the spy to have gone to some commanding position, nearer to the Greeks than Troy itself was, and I have already remarked that there is reason to believe that the ships were visible from Troy. Against its claims is to be placed the insignificance of the site, its unsuitableness for the position of a great city, and the improbability of its been chosen for the capital of the district; the want of correspondence with the description of the Homeric city; the absence of the important feature of two rivers joining their waters in front of it; and many other minor objections. It is almost surprising that such a situation should have found its advocates.

To turn now to the site at Hissarlik or Novum Ilium. Great

---

[60] 'Æn.,' xii. 742 foll.   [61] Ibid., i. 483.

[62] ἐν ᾗ νομίζεται τὸ παλαιὸν Ἴλιον ἱδρῦσθαι πρότερον (Strabo, xiii. i. 35). For the proof that Atchi-keui occupies the site of Pagus Iliensium I must refer the reader to Ulrichs' essay.   [63] Plato, 'Legg.' 682.   [64] 'Iliad,' ii. 791.

stress has been laid on the traditions in favour of this, but I have endeavoured to show that they were the inventions of a later period, and had no claims to real antiquity. Hellanicus, the historian, in the early half of the fifth century, is our first authority for the new city being supposed to have replaced the Homeric Ilium; and he went much further than this, and maintained that the inhabitants were lineal descendants of the old Trojans, and that these had originally come from Greece to Troy.[65] From this we may conclude that Strabo was not unfair in accusing him of doing it "to gratify the Ilians, as was his wont,"[66] for they were of Æolian origin, like himself: at all events, as there was no such thing as historical criticism in existence at that time, the statement proves nothing more than that it was believed by the inhabitants. More importance is to be attached to the fact that the Dumbrek was called Simois in the time of Demetrius of Scepsis; but there is reason for believing that almost all the features of the district were renamed subsequently to the Homeric period.[67] On the other hand, the following considerations, several of which are adduced by Strabo, seem fatal to its claims: (1.) The distance from the sea was too short. If it was twelve stades in Strabo's time, it must have been less in Homer's: and even though we place the Greek ships at the furthest point from the city, it is impossible to conceive that they could have been left undefended when the enemy was so near, or that long continuous battles could have taken place in the intermediate space, or that there should have been danger of an ambush entering the city when the army was in the plain. (2.) There are no sources near it which could in any way correspond to those which Homer describes as rising in front of Troy, nor is there any reason for believing that there ever were such. (3.) There is no point which can represent the tumulus of Æsyetes, the look-out place of Polites, as that must have been nearer to the Trojan camp, and in a commanding position. The traditional site, according to Strabo, was five stades from New Ilium, in the direction of Alexandria Troas; that is, apparently, on the hills to the south of the city, and consequently in a most unsuitable place. (4.) The situation of the rivers is unlike what Homer describes. Welcker, who supposes that the Scamander flowed into the sea by Rhæteum, objects that the battle could not be described as "swaying to and fro in the plain between the Scamander and

---

[65] Hellanicus, *apud* Dion. Hal. i. 47, 48; quoted by Ulrichs, 'Reisen,' ii. p. 298.
[66] Strabo, xiii. i. 42.   [67] See above, p. 347, Note 40.

Simois," since they would flow at right angles to one another; and that the Simoeisian and Scamandrian plains would then be in separate valleys, and would not form one plain, as is implied in the 'Iliad.'[68] The latter of these objections loses its force, if we suppose the Scamander to have flowed by Sigeum, for then both rivers would run for some distance through an open plain, in front of the city; but as this disappears, other and more formidable ones arise. For if, on this supposition, the ships are to be placed between the river and the Haven of the Achæans, the Simois, and not the Scamander, will intervene between them and the city; if, on the other hand, they are supposed to stand on the left bank towards Sigeum, either both streams must be crossed, or a long détour must be made by the Greek army at the foot of the hills, before the real fighting-ground can be reached. None of these suppositions are reconcilable with the Homeric narrative.

I have already stated my reasons for believing that Troy stood on the hill above Bunarbashi, in comparing the Homeric description with the present features of the ground.[69] Let me again briefly enumerate them. This site has fountains rising at its foot in the position which Homer describes. In front of it flow two rivers, for some distance parallel to one another, then joining their waters and flowing into the Hellespont. It is lofty, craggy, and very conspicuous, being exactly the situation which the inhabitants of these countries preferred for their cities, and superior to any other in the neighbourhood. It commands the entire length of the plain, facing the shore where the Greeks are supposed to be encamped. Behind it lies a small plain, near the river, and in the direction of Ida, corresponding to the Ileian plain. In front is a hillock, which may be supposed to be that called Batiæa, and on one side, at some distance off, and nearer the Hellespont, a commanding look-out station on the tumulus called Ujek-tepe. It is also to be observed generally concerning the tumuli in this district, that they are found all about the sides of the plain, and at its head—not merely about the lower part, as we should expect, if Novum Ilium and its vicinity were the scene of the 'Iliad.'

It now remains to notice the objections to this site; they are as follows:—

(1.) Zeus, when seated on the summit of Ida, is represented as "looking at the city of the Trojans and the ships of the Greeks."[70] Now, though some of the lower heights of the chain of Ida are visible from the Balidagh, the highest point, or Gargarus, is certainly not so.

---

[68] Welcker, p. xlv. [69] See above, vol. i. pp. 39 foll. [70] 'Iliad,' viii. 47-53.

It might be thought a sufficient answer that this was indifferent to the god's all-seeing eye, and that he is described as viewing the same objects from Olympus, which is not seen at all from Troy. But when Zeus descends to Ida with the express object of getting a nearer view of the conflict, we cannot help expecting that what he comes to see will be actually visible. The true reply probably is, that what is here intended by "the city and the ships" is the battle-field of which they were the two boundaries: and this is corroborated by what we find in other passages. Thus, when Iris is sent from Ida to the plain to order Poseidon to cease from the conflict, she is described as going to "sacred Ilium";[71] and when Hecuba desires Priam to offer a libation to Zeus, "who overlooks all Troy,"[72] she seems to mean the district as well as the city.

(2.) Great exception has been taken to the distance of Bunarbashi from the sea, which is thought to be much greater than what is implied in the 'Iliad.' In particular it is said, Homer would not have represented the armies as chasing one another up and down a plain of such length several times in one day. But we must remember that if the shore line was not so far advanced in Homer's times, the distance would be less than at present; and that the persons introduced on the scene are not ordinary mortals, but men of the heroic age: besides which, something may fairly be conceded to poetical exaggeration. There are, however, other passages, to some of which I have already referred,[73] that imply that the city and the Greek camp were near one another. In these the difficulty belongs to the poem itself, not to the locality; for what is required is to reconcile them with another set of passages, which not less certainly prove that they were distant from one another. On this subject we may be content to think with Colonel Leake that "at one time the poet found it convenient to magnify beyond probability, or even beyond possibility, the common occurrences of war; at another, to bring together the actions of an extensive field, in order to present them to view in one continued scene."[74] But, while it is possible for a poet to foreshorten an extensive tract of ground, to expand a narrow one, like that between Novum Ilium and the sea, is not within the limits of poetic license.

In conclusion, though the view which has been advocated cannot be

---

[71] 'Iliad,' xv. 169.   [72] Ibid., xxiv. 291.   [73] See above, vol. i. pp. 24, 25.
[74] 'Tour in Asia Minor,' p. 303.

regarded as certain, and though in some particular points we might wish to discover a closer resemblance between the features of the ground and the Homeric descriptions, yet I would submit that the correspondence between the two has been fairly made out, and that we have determined as much as we can expect to do, when we consider that the only authority to guide us is a very ancient document, and that document a poem.

APPENDIX B.[1]

GREEK INSCRIPTION FOUND AT MONASTIR.

THE inscription, of which a facsimile is given on the opposite page, now exists at Monastir, and was brought thither from an ancient site, believed to be that of Deuriopus, twelve miles distant from that city. The following is a cursive version :—

Παρὰ Φιλίππου τοῦ Ποσιδίππου, Ὀρέστου τοῦ Ὀρέστου, Παραμύθου τοῦ Μ[ε·]κίνου, τῶν ἀποκληρωθέντων πρ[ο]έδρων, δόγματος ἀναγραφή. Τῇ τ' τοῦ Δαισίου μηνός, τοῦ γμς' ἔτους· τῶν περὶ Βλ . . ὃρον Φιλίππου ἐν Δερμιόπῳ πολιταρχῶν συναγαγόντων τὸ βουλευτήριον, καὶ Φίλωνος τοῦ Κόνωνος ποιησαμένου λόγους περ[ὶ] Οὐεττίου Φίλωνος τοῦ θείου, καὶ προσαγγείλαντος ὅτι καὶ πρῶν? τὴν ἑαυτοῦ πατρίδα ἐτείμησε μεγάλως καὶ τελευτῶν οὐδὲ τῆς κατὰ τὴν βουλὴν τε[ι]μῆς ἠμέλησεν ἀλλ' ἀφῆκεν αὐτῇ κατὰ διαθήκην ⌒ αφ' < ἐφ' ᾧ ἐκ τῶν κατ'ἐνιαυτὸν ἐξ αὐτῶν γεινομένων τόκων ἡμέραν ἄγουσα Οὐεττίου Βωλάνου ἑορτάσιμον εὐωχῆται τῇ προδεκατεσσάρων καλανδῶν Νοεμβρίων· ἔδοξεν τῇ Βουλῇ τὴν τοῦ ἀνδρὸς σεμνότητα καὶ βούλησιν ἀποδέξασθαι, ἐπί τε ταῖς ὑπ' αὐτοῦ κατὰ τὴν διαθήκην γεγραμμέναις αἱρέσεσιν τό τ' ἀργύριον λαβεῖν καὶ κατ'ἐνιαυτὸν ἄγειν τὴν τοῦ Οὐεττίου Βωλάνου ἑορτάσιμον ἐκ τῶν τόκων ἡμέραν, καὶ μήτε τοῦ προγεγραμμένου κεφαλαίου ἀπαναλίσκειν τι εἰς ἑτέραν χρείαν μήτε τοῦ κατ' ἐνιαυτὸν γινομένου τόκου, ἀλλ' ὡς ὁ δοὺς Φίλων ἠθέλησεν. τό τ' ἀργύριον ἠριθμήσατο καὶ παρέλαβεν ὁ ἐπιμελητὴς τῶν τῆς βουλῆς δηναρίων Λούκιος Λουκρήτιος Πούδης.

l. 3. ΠΡΩΕΔΡΩΝ is a mistake of the sculptor or the transcriber for ΠΡΟΕΔΡΩΝ. τῇ τ'*1* sc. ἡμέρᾳ : probably the 19th ; *see* Franz, 'Elementa Epigraph. Græc.,' p. 349. The small mark after the T is one of the sigla which it was customary in inscriptions of this date to put after, or even before and after numeral letters: cf. l. 12, and Franz, 'Elem. Epig. Græc.,' p. 375.

l. 4. The Macedonian month Dæsius corresponded with the Attic month Thargelion. The numeral letters dating the year are to be

---

[1] This inscription has been edited for me by my friend, Mr. E. L. Hicks, Fellow and Tutor of Corpus Christi College, Oxford.

ΠΑΡΑ ΦΙΛΙΠΠΟΥ ΤΟΥ ΠΟΣΙΔΙΠΠΟΥ ΟΡΕΣΤΟΥ
ΤΟΥ ΟΡΕΣΤΟΥ ΤΡΑΜΘΥ ΤΟΥ ΜΚΝΟΥ ΤΩΝ ΑΠΟΚΛΗ
ΡΩΘΕΝΤΩΝ ΡΩΕΔΡΩΝ ΔΟΓΜΤΟΣ ΑΝ ΒΑΦΗ ͲΤ 1
ΤΟΥ ΔΑΙΣΙΟΥ ΜΗΝΟΣ ΤΟΥ Γ ΜΣΕ ΤΟΥΣ ΤΩΝ
ΠΕΡΙΒΛΓΕΔΡΟΝ ΦΙΛΙΠΠΟΥ ΕΝ ΔΕ ΡΡΙΟΠΩ ΠΟΛΙ
ΤΑΡΧΩΝ ΣΥΝΑΓΑΓΟΝΤΩΝ ΤΟ ΒΟΥΛΕΥ ΤͰΡΙΟΝ
Κ ΦΙΛΩΝΟΣ ΤΟΥ ΚοΝΩΝΟΣ ΠΟͰΣΑΜΕΝΟΥ ΛΟ
ΓΟΥ ΣΠΕΡ Ν ΟΥΕΤΤΙΟΥ ΦΙΛΩΝΟΣ ΤΟΥ ΘΕΙΟΥ
Κ ΠΡΟΣΑΝΓΕΙΛΑΝΤΟΣ ΟΤΙ Κ Ε ΩΝͰΛΕΑΥ ΤΟΥ ΠΑ
ΡΔΑΕ ΤΕΙ ΜͰΣΕ ΜΕΓΑΛΩΣ Κ ΤΕΛΕΥΤΩΝ ΟΥΔΕͰΣ
ΚΑΤΑ ͰΝ ΒΟΥΛͰΝ Μ ΙΣ ΗΜΕΛΗΣ ΕΝΑΛΛΑ ΦΙΕΝ
ΑΥ Ͱ ΚΑΤΑ ΔΙΑΘΗΚΗΝ Δ ΑΦ ΕΦΩ ΕΚ ΤΩΝ ΚΑΤΕ
ΝΙΑΥΤΟΝ ΕΞ ΑΥΤΩΝ ΓΕΙΝΟΜΕΝΩΝ   ΤΟΚΩΝ
ΗΜΕΡΑΝ ΑΓΟΥΣΑ ΟΥΕΤΤΙΟΥ ΒΩΛΑΝΟΥ ΕΟΡΤΑ
ΣΙΜΟΝ ΕΥΩΧΗΤΑΙ ͰΠΡΟ ΔΕΚΑΤΕΣΣΑΡΩΝ
ΚΑΛΑΝΔΩΝ ΝΟΕΜΒΡΙΩΝ ΕΔΟΞΕΝ Ͱ ΒΟΥΛΗ
ΤΗΝ ΤΟΥ ΑΝΔΡΟΣ ΣΕΜΝΟΤΗΤΑ Κ ΒΟΥΛΗΣΙΝ
ΑΠΟΔΕΞΑΣΟΛΙ ΕΠΙΤΕΤΑΙΣ ΥΠ ΑΥΤΟΥ ΚΑΤΑ
ΤΗΝ ΔΙΑΘΗΚΗΝ ΓΕΓΡΑΜΜΕΝΑΙΣ ΑΙΡΕΣΕΣΙΝ
ΤΟ ΑΡΓΥΡΙΟΝ ΛΑΒΕΙΝ ΚΑΙ ΚΑΤ ΕΝΙΑΥΤΟΝ ΑΓΕΙΝ
ΤΗΝ ΤΟΥ ΟΥΕΤΤΙΟΥ ΒΩΛΑΝΟΥ ΕΟΡΤΑΣΙΜΟΝ ΕΚ
ΤΩΝ ΤΟΚΩΝ ΗΜΕΡΑΝ ΚΑΙ ΜΗΤΕ ΤΟΥ ΠΡΟΓΕΓΡΑΜΜΕ
ΝΟΥ ΚΕΦΑΛΑΙΟΥ ΑΠΑΝΑΛΙΣΚΕΙΝ ΤΙ ΕΙΣ ΕΤΕΡΑΝ
ΧΡΕΙΑΝ ΜΗΤΕ ΤΟΥ ΚΑΤ ΕΝΙΑΥΤΟΝ ΓΙΝΟΜΕΝΟΥ ΤΟ
ΚΟΤ ΑΛΛΩΣ Ο ΔΟΥΣ ΦΙΛΩΝ ΗΘΕΛΗΣΕΝ ΤΟ ΑΡ
ΓΥΡΙΟΝ ΗΡΙΘΜΗΣΑΤΟ ΚΑΙ ΠΑΡΕΛΑΒΕΝ Ο ΕΠΙΜΕΛΗ
ΤΗΣ ΤΩΝ ΤΗΣ ΒΟΥΛΗΣ ΔΗΝΑΡΙΩΝ ΛΟΥΚΙΟΣ
ΛΟΥΚΡΗΤΙΟΣ ΠΟΥΔΗΣ

read backwards, σμγ´, cf. Boeckh 'Corp. Insc.,' 1062; the number will then be 243. The probable era from which this date is calculated is the Æra Achaica, B.C. 146, when Greece was made a Roman province: see 'Corpus Inscr.,' 1053, 1062. This would make the date of this inscription A.D. 97. But it is not easy to make sure in all cases on what era a date is based.

l. 5. The words from **ΤΩΝΠΕΡΙ** to **ΦΙΛΙΠΠΟΥ** are very difficult to make out. In one letter at least the copy is defective, and probably the obscurity is increased by contractions. Possibly the word ΒΛ . . . . . ΔΡΟΝ may be a proper name, ΦΙΛΙΠΠΟΥ being added as the father's name. In that case τῶν might perhaps be taken with πολιταρχῶν, the meaning of τῶν περὶ Βλ . . . δρων Φιλίππου ἐν Δεῤῥιώπῳ πολιταρχῶν being "the colleagues of Bl . . . der, Politarchs at Derriopus." But it is very doubtful.

Δεῤῥιώπῳ. This name is spelt Deuriopus in Strabo and Livy. Stephanus Byzantinus calls it Δουρίοπος. He is also the only authority for this being (as here) the name of a city as well as of a territory. Δουρίοπος, πόλις Μακεδονίας· Στράβων ἑβδόμῃ. οὕτω καὶ ἡ χώρα.

Πολιταρχῶν. As this name has not before been found as the title of the officers of a city except at Thessalonica, its occurrence here is interesting, as showing that it was in use elsewhere in Macedonia. There are two forms of the word, in ος and ης, but the title in the Acts is πολιτάρχης; the same variation is found in ἑκατόνταρχος, ἑκατοντάρχης. See also vol. i. p. 145.

l. 8. ΠΕΡΝ, a mistake of the sculptor or transcriber for ΠΕΡΙ. Translate, "concerning Vettius, uncle of the said Philo."

l. 9. If I have read the copy aright, which is here very indistinct, πρῶν must be a contraction for πρώην, in the sense of "formerly," "previously to his death." A similar form is quoted by L. and S. from Callimachus. We should perhaps have expected ζῶν.

l. 12. I think I have read the copy aright, although it is here very faint. For the sigla which enclose the numeral letters, see note on l. 3. If ΛΦ be the reading, it will amount to 501, the letters being read backwards as in l. 4. The silver denarius was the usual coin of reckoning at this time; this would make the sum bequeathed by Vettius Bolanus something between 17*l.* and 18*l.* Even assuming a high rate, the yearly interest would scarcely be more than 2*l.* Although the precious metals were very scarce in Greece at this time (see Finlay's 'Greece under the Romans,' p. 88), so that the rate of interest would be high, and the purchasing power of money considerable, yet this can hardly be thought a sufficient sum to feast a Boulê. Either, therefore, the transcriber is at fault, or I have misread him.

l. 15. In Latin it would be A.D. xiv. Kal. Novemb.

l. 19. αἱρέσεσιν, "intentions" expressed in the will; *i.e.* "conditions." Cf. προαίρεσις.

l. 25. ὁ δοὺς Φίλων. Philo appears to have been the nephew and executor of Vettius Bolanus. Cf. l. 8.

l. 27. δηναρίων. The silver denarius was the usual money of reckoning at this period. Cf. note on l. 12.

l. 28. In Latin, "Lucius Lucretius Pudens."

APPENDIX C.

ON THE RETREAT OF BRASIDAS FROM LYNCESTIS.

The mountain-side which has been described as intervening between Gurnitzovo and Tulbeli was most probably the scene of the remarkable retreat effected by Brasidas, the Spartan general, in the ninth year of the Peloponnesian war. During his campaign in Chalcidice, that commander was in alliance with Perdiccas, the Macedonian king; and in order to retain his good-will he was forced, on two separate occasions, to join him in invading Lyncestis—that is, the district at the southern end of the plain of Monastir. In the latter of these invasions their combined forces had penetrated into that country and defeated Arrhibæus, the king, after which they remained stationary for some time, waiting the arrival of some Illyrian mercenaries who had agreed to come to their assistance. After some days the Illyrians appeared, having crossed the intervening mountains to the west; but they broke faith with Perdiccas and joined his adversary, in whose favour their numbers enabled them to turn the scale. When the news of this reached the Macedonian camp the soldiers were seized with panic, and fled homewards towards Edessa in confusion during the night. Brasidas, who was encamped at some little distance off, was unaware of what had taken place, and the next morning found himself deserted by his allies, and Arrhibæus with the Illyrians close upon him; he had consequently no choice left but to make the best of his retreat. In this he succeeded without difficulty as long as they were in the plain; for the Greeks being trained soldiers, while the Illyrians were undisciplined barbarians, he was able to repel them with loss whenever they approached to the attack. Finding themselves thus rudely handled, the barbarians ceased from molesting them in the plain, and, running forward, occupied the mountains by which the invaders would have to return towards the lower country. Consequently, when Brasidas arrived there he found the summit of the pass defended and the heights on either side of it already seized by his opponents, while others were moving round to take him in the rear. From this position it was necessary to dislodge them. Accordingly, he gave orders to three hundred of his best soldiers to charge up the more accessible of the two hills, and dislodge that part of the enemy before they should be hemmed in by their superior numbers. This attack succeeded; and now that the Lacedæmonians held one of the heights commanding the pass, the barbarians were forced to retire from that also, and leave the passage free for the army to cross. Moreover, the boldness of their enemies had inspired them with such fear that they did not further molest them in their retreat, and Brasidas was enabled to reach the territory of Perdiccas before nightfall.[1]

The passage here spoken of is called by Thucydides "the defile of Lyncus,"[2] and by Polybius, who mentions it in connection with another engagement, "the mountain-pass leading into Eordæa,"[3] that country being the district intervening between Lyncestis and Lower Macedonia: in other words, the neighbourhood of the lakes of Ostrovo and Sarigöl. The position of this corresponds well with the pass by Gurnitzovo; and as the Egnatian Way at a later time appears to have followed the same route on its way to Edessa,[4] it would seem to have been the natural line of communication. The only difficulty is that the mountain-slopes on this side of the pass are more open than the historian's description would lead us to suppose, although about the summit there are heights which might easily be defended. The ascent would be a long one, as it took us two hours and twenty minutes at a foot's-pace to descend.

---

[1] Thuc., iv. 124-8.　　[2] Ibid., iv. 83.
[3] Polyb., xviii. 6.　　[4] Leake, 'Northern Greece,' iii. p. 317.

APPENDIX D.

ON THE EGNATIAN WAY.

As Elbassan is the point where we leave the Via Egnatia—the course of which we have followed from Salonica in Chapters vii.-ix.—it may be well to look back from that point over the route we have taken, and to see what can be determined with regard to that important line of ancient communication. The name Egnatia was applied to the entire length of the road from the Adriatic to Constantinople; while the western half, extending as far as Thessalonica, was sometimes called, for the sake of distinction, the Via Candavia: it is with the latter of these alone that we have now to do. This road started from two separate points on the Adriatic—Dyrrachium and Apollonia—and the branch-lines from those cities converged at a place called Clodiana; from thence it followed the valley of the Genusus for some distance, and then, penetrating through the Candavian mountains, and passing the northern shore of the Lacus Lychnitis, reached the important town of Lychnidus. Between this city and Heraclea the passes of the Scardus had to be crossed; and from the latter place the road continued through Lyncestis and Eordæa to Edessa, and thence by Pella to Thessalonica. Our authorities on the subject are Strabo, who tells us the entire length of the Via, and gives a general description of the country through which it passes; and the Itineraries, which give us the names of the different stations along the line and the distances in Roman miles between them. Of these, the Antonine Itinerary contains two separate enumerations—one starting from Dyrrachium, the other from Apollonia—which present considerable variations, both in names and numbers, in the part that is common to both. The Jerusalem Itinerary also, and the Tabular Itinerary, or Peutinger Table, contain this line of road, though they too differ in many points both from one another and from the Antonine. The subject has not hitherto been satisfactorily investigated: and, in particular, none of the towns that are mentioned can be certainly determined westward of Edessa (Vodena). Yet, as Colonel Leake has said, until some of the ancient sites have been ascertained, no safe criticism can be exercised on the Itineraries themselves. Leake himself did not penetrate westward of the Scardus, and therefore, though his suggestions are characterised by his usual good judgment, they cannot be considered to determine anything. The principal authority on the subject is Tafel, in his work, 'De Viâ militari Romanâ Egnatiâ,' a work of great learning and permanent value, as it contains an examination of all the passages that in any way elucidate it in ancient, mediæval, and modern writers; but the author has never himself visited the country, and appears to be somewhat deficient in geographical insight. The consequence of this is, that in the western part of the route almost all his conclusions seem to be erroneous.

The entire length of the Via from Dyrrachium to Thessalonica is given by Strabo[1] as 267 miles, and with this the numbers in the Itineraries fairly agree: the Antonine (ed. Parthey and Pinder) giving the total as 269, and the Tabular as 275; while the Jerusalem gives the distance from Thessalonica to Apollonia as 303, which would make that from the former place to Dyrrachium to be 287, or perhaps less, as Clodiana, the point of junction, was at least 16 miles further from Apollonia than from that city. Strabo is certainly in error when he says, that the distance was the same whether the traveller started from Apollonia or from Dyrrachium.

The following are the numbers in the Itineraries which refer to this route, and the modern distances computed by hours:—

ANTONINE I.

Dyrrhachium.   M.P.
Clodiana   33
Scampis   20
Tres Tabernas   28
Lignido   27
Nicia   32
Heraclea   11
Cellis   34
Edessa   28
Pella   28
Thessalonica   28

ANTONINE II.

Apollonia.
Ad Novas   24
Clodianis   25
Scampis   22
Tribus Tabernis   30

ANTONINE II.—continued.
                       M.P.
Lignido   27
Scirtiana   27
Castra   15
Heraclea   12
Cellis   33
Edessa   33
Diocletianopolis   30
Thessalonica   29

JERUSALEM.

Thessalonica.
Mutatio ad decimum   10
Mutatio Gephyra   10
Civitas Pelli   10
Mutatio Scurio   15
Civitas Edessa   15
Mutatio ad duodecimum   ..   12

---

[1] Strabo, vii. 7, § 4.

JERUSALEM—continued.

	M.P.
Mansio Cellis	16
Mutatio Grande	14
Mutatio Melitonus	14
Civitas Heraclea	13
Mutatio Parambole	12
Mutatio Brucida	19
Civitas Cledo	13
Mutatio Patras	12
Mansio Claudanon	4
Mutatio in Tabernas	9
Mansio Grandavia	9
Mutatio Trajecto	9
Mansio Hiscampis	9
Mutatio ad Quintum	6
Mansio Coladiana	15
Mansio Marusio	13
Mansio Absos	14
Mutatio Stephanophana	12
Civitas Apollonia	18

PEUTINGER.[2]

Dyrrachium.

	M.P.
Clodiana	26
Scampis	20
Genusus fl.	9
Ad Dianam	7
Candavia	9
Pons Servili	9

PEUTINGER—continued.

	M.P.
Lignido	19
Nicea	16
Heraclea	11
Cellis	32
Edessa	45
Pella	45
Thessalonica	27

MODERN DISTANCES COMPUTED BY HOURS.[3]

Salonica.

	HRS.	MIN.
Vardar Khan	4	15
Pel	4	0
Yenidje	1	20
Vodena	6	20
Ostrovo	4	0
Gurnitzovo	3	30
Tulbeli	2	20
Monastir	5	45
Resna	6	5
Ochrida	5	5
Struga	2	30
Kukus	7	0
Skumbi-bridge	8	0
Elbassan	3	0
Pekin	7	0
Durazzo	8	0

[2] There is, unfortunately, great difficulty in knowing to which stages the numbers apply in this portion of the Table, and they are taken differently by Tafel on the one side, and Leake and Hahn on the other (Tafel, p. 4; Leake, 'N. Greece,' iii. p. 313; Hahn, 'Reise von Belgrad,' p. 236). Between Lychnidus and Nicæa the road breaks off, and a stage apparently is omitted. Hahn's ingenious emendation of the almost impossible numbers xlv. on either side of Edessa is worthy of notice. He suggests that as Vodena is halfway between Monastir and Salonica, these numbers originally signified that Edessa was 45 miles distant both from Heraclea and from Thessalonica, and that in this manner they were gradually introduced by transcribers into their present position in the Table, in place of the previously existing numbers for those stages, 13 and 18.

[3] N.B.—The hour varies from 2¾ or 3 miles to 4 miles; the former being

In endeavouring to determine the position of the places mentioned in the Itineraries we meet with two considerable sources of difficulty. First, the distances in the Itineraries are computed by Roman miles, which represent a fixed measurement; whereas the only way in which they are now estimated is by hours, and the distance represented by an hour varies excessively according to the nature of the ground. Consequently, there is no certain standard by which to compare the two. Secondly, though the names of the principal stations and the distances between them correspond very fairly in the different Itineraries, yet in the minor stations which intervene between these there is great inaccuracy in the numbers and want of correspondence in the names. Thus, for instance, in the passes between Lychnidus and Heraclea, all the names are different in the two Antonine Itineraries and the Jerusalem—a circumstance which induces Leake to think that there must have been at one period a choice of routes over the mountain-ridges in that part.[4] Pons Servilii, an important position in a geographical point of view, is mentioned only in the Tabular. Candavia in the Jerusalem is 9 miles from Trajectus Genusi, while in the Tabular there are 16 miles between them, with an intermediate station called Ad Dianam; from which we might be inclined to suppose that the name of Candavia had at different times been applied to different places in the Candavian mountains. All the critics, in fact, agree that the Itineraries require considerable emendation; and the grotesque mistakes in the spelling of the names which we find (*e. g.* Cledo for Lychnido, Granda Via for Candavia, in the Jerusalem) render serious errors in the numbers only too probable. But simply conjectural emendation of numbers is almost a hopeless task.

The conclusion to which we are led by considering these difficulties is, that we should begin by endeavouring to determine those places the position of which seems to be fixed by the nature of the country. Fortunately, there are several of such a character. Fortunately, also, from the direction taken by the river-valleys and passes, the line of the road in most parts is almost absolutely determined beforehand.

Between Thessalonica and Edessa there is no difficulty about the line of road: the position of Pella is certainly found at the village

---

the pace in the mountainous districts, the latter on level ground. As far as Ochrida, the time is that of our own journey; the remaining part, as we travelled by the *menzil*, which is faster than the usual pace, I have taken from Boué ('Recueil d'Itineraires,' i. p. 267), and Hahn ('Albanesische Studien,' i. p. 134). [4] 'Northern Greece,' iii. pp. 311, 312.

of Alaclisi and the khan of Pel, and that of Edessa at Vodena. Mutatio Gephyra of the Jerusalem Itinerary evidently corresponds to the present bridge and ferry over the Vardar, allowing for some slight variation in the course of the stream.

To turn now to those places which are more or less determined by the topography of the country: the most important of these is the Trajectus Genusi, or crossing of the river Genusus, which occurs in the Jerusalem as Mutatio Trajecto, and in the Tabular as Genusus Flumen. This is almost certainly the same ford at the foot of the mountains by which the Skumbi is now crossed in ascending from Elbassan in the direction of Ochrida. Even if the name Candavia did not occur among the stations, we might be certain that the ancient road would ascend into the mountains on the south side of the stream, to avoid the considerable bend towards the north which it makes between this point and Kukus: and it must have crossed the river here from the right bank in order to avoid the narrow gorge from which it emerges. Having fixed this point, we can proceed to determine the site of Scampæ, which is placed in both those Itineraries 9 miles off from it in the direction of Dyrrachium: this position closely corresponds to the town of Elbassan. From that place to Durazzo the distance is estimated at 14 hours or 42 miles, which corresponds fairly with the numbers in the Itineraries. Clodiana, the station where the road to Apollonia diverged, was rather less than half-way from Scampæ to Dyrrachium, and was probably on the Genusus, as its name seems to have been derived from Appius Claudius, whose camp was on that river during the campaign against Gentius in B.C. 168.[5] Hahn believes it may be identified with a place called Pekin.[6]

The next point of importance is Pons Servilii, which lies some distance to the west of Lychnidus. There is a strong probability in favour of this being the bridge over the Drin at Struga, which must always have been an important position. The only other place between Lychnidus and Trajectus where there could have been a bridge is the crossing of the upper stream of the Skumbi near Kukus, and the river there is of no great size. Tafel,[7] indeed, prefers the latter position, and there is this in its favour, that it places Pons Servilii at a greater distance from Lychnidus, 19 miles being given as the distance between them in the Peutinger Table; whereas, if Pons Servilii were at Struga and Lychnidus at Ochrida, the distance would be only 9 miles. In order to rectify this, Hahn proposes to correct the Table by

---

[5] Livy, xliv. 30.   [6] 'Reise,' p. 237, *note*.   [7] 'De viâ Egnatiâ,' p. 31.

transposing this number 19 with the 9 of the previous stage from Candavia to Pons Servilii; thus subtracting 10 from the distance to Lychnidus, and adding on the same number in the direction of Candavia and the Trajectus. But, anyhow, if we examine the numbers, there proves to be considerably greater difficulty in placing Pons Servilii at Kukus than at Struga. For, if from the 19 miles between Lychnidus and Pons Servilii we deduct 9 for the distance between Ochrida and Struga, there remain only 10 for the stage between Struga and Kukus, which is computed as a journey of 7 hours. The geography of this part, however, is not as simple as Leake supposed it to be (see his map at the end of vol. iii. of 'Northern Greece'); for there is no direct route from Struga to the valley of the Skumbi, the only passage through the intervening mountains being at a point some way farther to the south, which involves a considerable détour.

The next point is a very important and a very difficult one,—the position of Lychnidus. Leake argues that if Pons Servilii was at Struga, as the distance from that place to Lychnidus is given in the Table at 17 (19?) miles, it follows that Lychnidus must have been at the south-eastern corner of the Lacus Lychnitis, and that the road must have crossed the Scardus range by a pass, or perhaps by several passes, eastward from that position, descending into the plain of Monastir at Florina, or, as he calls the place, Filurina.⁸ Now, even if we could trust the numbers in the Table, the 19 miles from Struga— *i. e.* 10 miles from Ochrida—would not bring us more than half-way down the eastern shore of the lake, which is upwards of 20 miles in length; but the real answer to Leake's view is, that the passes which he supposes do not exist; the only pass which leads through the mountains from Florina being that in the direction of Castoria, which is quite out of the line of the Egnatian Way. In fact, here again the nature of the ground comes to our assistance in determining the sites, for there is one and one only passage through this part of the Scardus range, namely, that which leads from Monastir to Ochrida. Now, the position of Ochrida, lying as it does near the foot of the pass, on the shores of the lake, must at all times have been an important one; and the solitary height on which its castle is built is so conspicuous a site, that it could hardly have been overlooked in ancient times. The probabilities, therefore, are strong in favour of Lychnidus, the chief town of the district, which was on the line of road, having been placed there.⁹

---

⁸ 'Northern Greece,' iii. pp. 281, 282.

⁹ The fact that the mile-stone mentioned in the text (vol. i. p. 198) was

The same argument from the position of the pass which seems to fix Lychnidus at Ochrida, applies with equal force to Monastir as the site of Heraclea; that place being situated exactly at the foot of the mountains on the opposite side. No doubt, the numbers given in the Itineraries between Lychnidus and Heraclea, though they vary amongst themselves, imply a somewhat longer distance than the 12 hours between Ochrida and Monastir; but, on the other hand, the distances on the ancient lines of road from Thessalonica and Stobi to Heraclea would place their point of convergence exactly at Monastir.[10] The arguments by which Leake has shown that Pelagonia was situated at Monastir are strongly in favour of Heraclea being placed there, and not, as he suggests, at Florina; for Tafel proves, from a passage of Cinnamus, that Heraclea and Pelagonia were two names for the same place.

It only now remains to speak of the road between Edessa and Heraclea. Cellæ, the principal station on the way, which is mentioned in all the Itineraries, may perhaps have been in the neighbourhood of Ostrovo; for the ancient road must have passed near that place, as there is only one route that it could have taken from Vodena as far as that point in the direction of Monastir.[11] There is great difficulty, however, in reconciling that position with the Itineraries, according to which Cellæ was nearly half-way from Edessa to Heraclea; whereas Ostrovo is not much more than one-quarter of the way from Vodena to Monastir. From that point the road would naturally cross the pass over the mountains by Gurnitzovo to Tulbeli, which offers the easiest access to the plain of Monastir.

---

found at Ochrida, bearing the inscription "eight miles from Lychnidus," cannot be adduced as an argument on the other side; for it does not exist *in situ*, but was brought to its present position from the castle, where it was found lying on its side. As heavy stones were useful for purposes of defence or for building, and water carriage along the lake was easy, it is quite natural that such a stone should have been removed in the course of the middle ages.

[10] *See* Hahn, 'Reise,' pp. 234, 236.

[11] Mr. Curtis, of Constantinople, informs me that he discovered last year, not far from the road at the summit of the pass between Vodena and Ostrovo, the remains of a Roman road, which was traceable for some distance in the direction of the head of the lake.

APPENDIX E.

THE BIRTHPLACE OF JUSTINIAN.

A VISIT to the city of Uskiub acquires a twofold interest when we consider that in its immediate neighbourhood was almost certainly the birthplace of the Emperor Justinian, who, notwithstanding the weakness of his character and the unsubstantial nature of his conquests, is always to be remembered as the originator of the greatest code of laws that has ever been framed, and as the builder of St. Sophia's. As great confusion has existed with regard to the place where he was born, and it is only within a year or two that the whole of the evidence on the subject has been laid before us, it may be worth while to say a few words on this point. For a long time Ochrida was considered to have been the fortunate spot. Great was my perplexity when, on returning to England after visiting that place, I looked at the name 'Achrida' in Le Quien's 'Oriens Christianus,' and found, along with various notices from mediæval writers of the foundation of the city and its position near the lake, an elaborate description of the buildings erected there by Justinian, when he gave it the name of Justiniana Prima, as having been his birthplace, and especially of an aqueduct, by which he brought a permanent supply of water into the place. This last fact at once staggered me, as I knew that the original city of Ochrida was situated on the summit of the castle-rock, to which no aqueduct could by any possibility be brought; so I had to mistrust my authority and look elsewhere for information. I found that Le Quien and other writers had been led astray by some of the Byzantine historians, as, for instance, Nicephorus Callistus and Nicephorus Gregoras, who have fallen into the same error in consequence of their finding that the metropolitans of Ochrida used in their signature the title of Archbishop of Justiniana Prima and Achridæ. Hence they inferred that these were two names for the same place, and that Achrida was Justinian's birthplace; whereas the more legitimate conclusion would be that two names would not have been used if one city alone had been intended. The explanation of the double title is, that while Justinian had established the metropolitan see at the place on which he bestowed his name, it was transferred to Ochrida when that city was made the capital of the mediæval Bulgarian kingdom, and the archbishops from that time forward employed both appellations. Even Colonel Leake, who is seldom caught napping, speaks of Justinian as having "founded at Achris the town which he named Justiniana Prima."[1] After this it is almost comical to hear Gibbon saying, in his grand style, "There is some difficulty in the date of his [Justinian's] birth; none in the place—the district Bederiana—the village Tauresium, which he afterwards decorated with his name and splendour."[2] The question is just this—where are we to look for Bederiana and Tauresium? The passage of Procopius to which Gibbon is referring, and which is our sole authority on the subject, runs as follows: "In the district of the European Dardani, who dwell beyond the confines of the Epidamnians, and close to a castle which bears the name of Bederiana, was a place called Tauresium, where the Emperor Justinian, the renovator of the world, was born. This spot accordingly he enclosed within a narrow space with walls in the form of a square, and placed a tower at each of the angles, from which circumstance it came to be called, what in fact it was, the Castle of Four Towers. Not far from the place he built a magnificent city, which he named Justiniana Prima, as an acknowledgment of what he owed to the place that reared him; though, indeed, it was a debt that the Romans, to a man, ought to have shared, seeing that he whom this land brought up was the common saviour of all. Here, too, he constructed an aqueduct, and benefited the city by providing a perennial supply of water, and carried out a number of other admirable works which confer great honour on the founder of the city. In fact, it is no easy task to enumerate the houses of God, the magistrates' residences, which pass description, the vast colonnades, the handsome squares, the fountains, streets, baths, and market-places. In a word, it is a large and populous city, prosperous in other respects, and worthy to be the capital of the whole district. For to this dignity it has been raised; and moreover it has been appointed as the seat of the Illyrian archbishop, and the other cities give way to it as taking precedence."[3] He goes on to say that Justiniana Secunda, from which it was distinguished, was the city of Ulpiana. Commenting on this passage a little further on in his history, Gibbon says, "The solitude of ancient cities was replenished; the new foundations of Justinian acquired, perhaps too hastily, the epithets of impregnable and populous, and the auspicious place of his own nativity attracted the grateful reverence of the vainest of princes. Under the name of *Justiniana*

---

[1] 'Northern Greece,' iii. 273.   [2] Smith's 'Gibbon,' v. 35.
[3] Procopius 'De Ædificiis,' iv. 1.

Prima the obscure village of Tauresium became the seat of an archbishop and a prefect, whose jurisdiction extended over seven warlike provinces of Illyricum; and the corrupt appellation of *Giustendil* still indicates, about twenty miles to the south of Sophia, the residence of a Turkish sanjak."[4]

The chief authority to whom Gibbon refers for the identification of these places is D'Anville, whose views are given in an Essay in the 'Mémoires de l'Académie.'[5] His arguments turn mainly on the similarity of name, and the general correspondence in the position ascribed to these cities; and in default of a better suggestion, they may be allowed to have considerable weight. At the same time Dardania, in which Procopius declares that Justiniana Prima was situated, seems never to have extended further east than Scupi, and Giustendil is not only some distance to the north-east of that place, but also too far removed from Ochrida for it to have been likely that the Metropolitan see should have been transferred from the one of these two cities to the other. Mannert[6] was the first, I believe, to point out that Uskiub, the ancient Scupi, was the only place that fulfilled all the conditions requisite to identify Justiniana Prima; and this view has been confirmed by subsequent discoveries, so that it may be now considered almost certain. It fell within the district of Dardania, and was situated at a moderate distance from Ochrida; it was also the most important position in that neighbourhood, and from having been the leading city would be most naturally pointed out for restoration and decoration. At Uskiub an aqueduct still remains, corresponding to that mentioned by Procopius; and Von Hahn, who passed by here in 1858, has shown that the names Tauresium and Bederiana may be traced in those of Taor and Bader, two villages lying near together in the neighbourhood of that city, the position of which has been noticed in the text.[7] If any objection arises to this from the fact that Procopius speaks of the building of a new town, and not of the restoration of an old one, it is sufficiently answered by his striking omission of Scupi in his enumeration of the towns which Justinian strengthened and fortified in this quarter: besides which, it almost follows that the new city must have been founded on a site already occupied, since we are told that it was large and populous, without any mention being made of a transplantation of the population from any other place. At the same time Procopius' motive is suffi-

---

[4] Gibbon, v. 78, 79.   [5] Vol. xxxi. pp. 287-292.
[6] 'Geographie der Griechen und Römer,' vii. 104 *seqq.*
[7] Hahn, 'Reise von Belgrad nach Salonik,' in the Proceedings of the Vienna 'Akademie der Wissenschaften' for 1860.

ciently evident, viz., to flatter the Emperor by representing him as the original founder of the city. It is likely enough that the scale on which this and similar restorations were carried out has been grossly exaggerated by the historian, and the old name of the place seems to have regained its ascendancy, except in official documents, not long after the death of Justinian.[s]

---

[s] In Smith's 'Dictionary of Geography,' owing, probably, to the plurality of contributors, some confusion seems to exist with regard to this point. Under the name *Justiniana Prima* the reader is referred to *Scupi*, as if the two were identical: but in the latter article there is no notice of Scupi having ever borne that name; and in the article *Lychnidus*, Justiniana Prima is identified with Giustendil.

APPENDIX F.

ON THE MARCH OF A ROMAN CONSUL ACROSS MOUNT OLYMPUS.

IN my account of the Lower Olympus, in Chapter XIX., I have dwelt with some minuteness on the features of the ground, because they illustrate a remarkable passage of a Roman army through the heart of this mountainous region in the year B.C. 169, of which Livy[1] has left us a detailed account; and as that historian derived his information from Polybius, who accompanied the army, his description is more than usually trustworthy.[2] The best part of M. Heuzey's volume is that which is devoted to the examination of this route, which he followed throughout; and I believe that his conclusions with regard to the various positions represent the truth. The circumstances described in Livy's narrative are as follows:—During the first years of the war with Perseus the Roman commanders had not advanced beyond Thessaly, and but little progress seemed to have been made. The Roman people appear to have been dissatisfied with this state of things, for early in the third year the consul Q. Marcius Philippus appeared on the scene with an entirely different plan of action, which soon changed the face of events. As soon as he had arrived in Thessaly, by way of Brundusium and Ambracia, he prepared at once to carry the war into Macedonia, and with that object entered Perrhæbia, as the district westward of Olympus was called, being as yet undecided by what route he should penetrate through the mountains. On hearing of his approach, Perseus occupied the passes; of which Tempe was guarded by a succession of forts; that called Volustana, leading through the Cambunian mountains some little distance to the west of Olympus, was held by a body of light armed troops; while that across Olympus itself was defended by a garrison of 12,000 men under Hippias, who were stationed at a fort on the further side of the lake Ascuris. No mention is made of the occupation of the pass of Pythium, the modern Petra, close under the north-west angle of the

---

[1] Livy, xliv. 1-9.
[2] Polyb. xxviii. 11, quoted by Leake, 'Northern Greece,' iii. 416, who has given a clear account of the campaign and its topography, as far as was possible without penetrating into the interior of the mountain.

mountain, but this was probably regarded as too narrow and too rugged to be attempted. The consul, who was stationed between Azorus and Doliche, two towns of Perrhæbia near the Titaresius, determined to attempt the route which here lay nearest to him, namely that leading through the heart of Olympus by Octolophus. Accordingly he sent forward a body of 4000 men under the command of his son, in conjunction with M. Claudius, to occupy the most favourable positions, and followed himself with the rest of the army. But the ground was so steep and rugged that at the end of two days the advanced party had only marched fifteen miles to a fort called Eudieron, which they captured, and the next day, proceeding seven miles further, occupied a height within sight of Hippias' position. The messenger who was sent back with tidings of these events to the consul found him already in the neighbourhood of the lake Ascuris, from which place he hastened forward to the support of the smaller force, and pitched his camp not much more than a mile from the enemy on the high ground already occupied. The situation of this is described as being so commanding, that the district of Pieria and the sea-coast between Dium and the exit of the Peneius were visible, together with the army of Perseus, and roused the ardour of the soldiers by holding out to them a prospect of bringing the war to a speedy conclusion.

Let us now compare the account hitherto given with the topography. The starting-point of the Romans between Azorus and Doliche was in the lower country westward of Sparmos, where sites corresponding to these cities have been found;[3] and Octolophus, in the direction of which the pass led, we have already identified with the mountain of Elassona, which forms a westerly continuation of the ridge on the southern side of the plain of Carya.[4] The ground, however, between that mountain and Skamnia is too steep to admit of the passage of a large force; and it seems more probable that they proceeded directly westward to the plain of Sparmos, from which the ascent is comparatively easy to Skamnia. Here would commence the passage into the plain of Carya over the shoulder of Mount Detnata, on the summit of which there are traces of an ancient site, which may represent Eudieron, as the distance corresponds to that given by Livy. Leake conjectured that that place might be found at Konospoli, in the plain of Carya, where he was told there were remains; but it appears that he was misinformed, for all that is to be seen there is an inscription on a boundary-stone which marked the line of separation between the

---

[3] Heuzey, 'Le Mont Olympe,' pp. 37 foll.
[4] See map on p. 17 of this volume.

territory of Dium and that of Oloosson. Descending to the plain they would cross it and approach the lake of Nezero (which is certainly the Palus Ascuris, for there is no other piece of water on Olympus) by the same route which we followed. As the whole mountain is intersected by the plain of Carya and the gorge of Kanalia, it may seem strange at first sight that they should not have selected this more direct passage to the plain of Pieria, for though that gorge is almost impassable for an army encumbered with baggage and elephants, yet we can hardly conceive anything worse than the place where they ultimately descended. But the truth seems to be that they intended to have followed the ordinary route which led downwards by an easy declivity to the country at the mouth of the Peneius; and perhaps also they were afraid of leaving Hippias in their rear. Lapathus, the place where that officer was stationed, is probably Rapsani, a village high up in the mountains overlooking Tempe; for one of the forts which defended that pass is expressly mentioned by Livy as being near Lapathus, and the modern name sounds like a corruption of that name. Consequently, when Livy speaks of it as being "super Ascuridem paludem," we must understand him to mean *beyond* and not *above* the lake; and this interpretation is confirmed by the statement that the advanced guard, when they approached that place, had to send a messenger to the consul who was in the neighbourhood of the lake, implying that there was some distance between them. The position of the Romans is placed by Heuzey on Mount Livadaki, or, as he calls it, Metamorphosis, by the side of the lake; and that summit is certainly the most commanding one in the whole district; but its distance from Rapsani, or any point which could represent Lapathus, is too great to be reconcilable with the historian's distinct statement of the proximity of the two forces, so that it should probably be placed on one of the heights near Rapsani, which command a view corresponding to that which is described. At the same time it is impossible to make this agree with the short space of seven Roman miles which is given as the interval between Eudieron and the station of the Romans.

Let us now return once more to the narrative. The day following his arrival in front of the enemy the consul Marcius devoted to resting his army, wearied by their long march; but on the two subsequent days there were engagements between the opposed forces, though from the difficult nature of the ground only the light troops could take part in them. It was impossible, however, for the Roman general to remain where he was, and accordingly he ventured on the bold step of leaving Hippias in his rear and descending at once to the plain; this he could hardly have done, had he not known what sort of an opponent he had to deal with in Perseus, but, as it was, that commander neither reinforced Hippias, nor attempted to harass the consul's march. Popilius was left on the heights to watch the enemy near Lapathus; and Marcius, being excluded from the usual descent towards Phila, at the eastern end of Tempe, was forced to explore for himself a passage further to the north-east, where he descended by a track under ordinary circumstances impassable to an army. On the second day he was joined by Popilius, when the whole force pressed onward by the pass of Callipeuce, and, after four days of unremitting labour, occupied in a descent second only to that of Hannibal in crossing the Alps, they pitched their camp in the plains between Heracleium and Libethrium. In this account there is little that requires explanation. As Heracleium is certainly Platamona, and Libethrium probably Leftocarya, the passage must have been made some miles to the south of the gorge of Kanalia; and the pass of Callipeuce would be one of the wooded glens which lead from the neighbourhood of Nezero and Rapsani towards the plain in that direction. The daring move of the consul Marcius had its immediate effect. Perseus was thunderstruck, and whereas it was still in his power to reduce the invaders to great straits, enclosed as they were between the mountains, the sea, and the enemy's strongholds, he immediately withdrew his garrisons from Tempe, and even abandoned Dium at the approach of the Romans. The only place in the whole of Pieria that made any resistance was Heracleium, which from the strength of its position held out for some time, until it was at last taken by assault, the besiegers mounting on a *testudo* formed by the compacted shields of the other soldiers. The remainder of this year's campaign was unimportant and undecisive; but it paved the way for the great victory of Æmilius Paullus at Pydna in the following year, by which the power of Perseus was shattered, and Macedonia reduced for ever into subjection to Rome.

APPENDIX G.

ON THE SITE OF DODONA.

DODONA is spoken of by classical writers sometimes as belonging to the district of Thesprotia, sometimes to that of Molossis. This Strabo had remarked, for whilst he himself places it in Molossis, he notices that the tragic poets call it a Thesprotian town. And Æschylus, though he speaks of the oracle as dedicated to Thesprotian Zeus, yet places it in the Molossian plains. Now the meeting-point of these two districts must have been somewhere in the close vicinity of the lake Pambotis (as the lake of Yanina was called in ancient times), though that piece of water was regarded as being actually in Molossis. Again, as Pindar describes Epirus as beginning at Dodona, and extending from thence to the Ionian Sea, it follows that Dodona must have been on the eastern frontier of Epirus. The distances also from the coast suit this position very well; Dodona is spoken of as being two days' journey from Ambracia (Arta), and four days' from Buthrotum (Butrinto): and though the distances along the two routes do not differ very greatly from one another, as seen on the map, yet the former, lying through a river valley, can easily be traversed in the time mentioned; while the latter, being in the heart of a wild and mountainous region, would probably require the longer period.

The only description of the neighbourhood of Dodona which has come down to us from ancient times occurs in a fragment of Hesiod, where the following account is given : " Hellopia," the poet says, " was a country of corn-fields and meadows, abounding in sheep and oxen, and inhabited by numerous shepherds and keepers of cattle, where on an extremity stood Dodona, beloved by Jupiter; here the God established his oracle in a wood of ilex, and here men received responses, when bearing gifts and encouraged by favourable omens they interrogated the god." The pastoral character of the country and its occupants here described is singularly in accordance with the abundant meadows which extend throughout the whole valley of Yanina, but are otherwise uncommon in so rocky a country as Albania. It is, perhaps, surprising that no mention is made of the lake, but there is some evidence to show that such a piece of water existed near Dodona, for Strabo implies that there were marshes near the temple; and whilst one ancient author speaks of Neoptolemus, the son of Achilles, as having settled on the shore of the lake Pambotis in Molossia, Pindar, who follows the same legend, considered Dodona as forming part of the domain of Neoptolemus. We learn also that it was situated under a high mountain called Tomarus, from the foot of which issued a hundred fountains. This could not be Mount Tomohr near Berat, on account of the position assigned to it; and it is probable enough that it was Mount Metzikeli, from the base of which the numerous sources emerge which contribute in great measure to supply the lake. The name itself seems to be preserved in the Tomarochoria, as some villages are called on part of the northern extremity of Mount Drysco, which is a continuation of Metzikeli.

Putting all the evidences together, we seem to be led with great likelihood to the conclusion that Dodona stood near this lake; and when we take into account the good taste which the Greeks always evinced in the position of their sacred edifices, and their belief that the gods delighted in places rendered remarkable by natural causes, which is amply evidenced by Delphi, Lebadea, and numerous other shrines, it seems highly probable that the oracle itself occupied the striking rocky promontory which here projects into the water. Still, at present an element of uncertainty remains; but as there can hardly fail to have been numerous inscriptions in the neighbouring country relating to so famous a place, there is every reason to hope, that when these wild districts have been more fully explored, we shall no longer be left in doubt as to this, the only place of great celebrity in Greece of which the situation is not exactly known.[1]

---

[1] The arguments here adduced are given more at length in Leake, 'Northern Greece,' vol. iv. pp. 168 foll., where also the references to the authors named above may be found.

INDEX.

A.

Abdurrahman, Pasha, visit to, i. 223.
Acanthus, city of, now Erisso, i. 128.
Acheron River, ii. 208, 218. Gorge of, 216.
Achmet Bey, i. 9.
Achrida, ancient, i. 186.
Acland, Dr., on the Plains of Troy, i. 40.
Acté, ancient, i. 54.
Adramyttium Bay, i. 16.
Æsyetes, tumulus of, look-out station of the Trojans, i. 40.
Aghia, town and plain of, ii. 79.
Alaklisi, village of, i. 154.
Albania, historical heroes of, i. 214. Southern, ii. 184.
Albanians, the, i. 209. Their language, 210. Character, 211. Superstitions, 212. Former condition, 220.
Albanian Beys, massacre of, i. 167. Riddles, 211.
Alexander the Great, i. 154. Birthplace of, ib.
Alexandria Troas, Roman remains at, i. 21.
Ali Pasha, his birthplace, i. 227. Appropriation of villages, 163. Measures against the Clefts, ii. 54. His tomb, 193. Sketch of his character by Mr. Finlay, 194. His massacre of Greek women, 197. Expulsion of the Suliotes, 215.
Alipuchori, village of, ii. 203.

Ambelakia, ii. 62. Trade in dyed thread, 63. M. Beaujour's account of its former mercantile prosperity, 63. Its decline, 65.
Anaurus, ancient, ii. 128.
Anthimus, ex-patriarch of Constantinople, i. 129.
'Arabian Nights,' The, ii. 278.
Arachova, Ulrichs' description of, ii. 230.
Archimandrite, i. 74.
Argyro-Castro, town of, i. 228. Its population, 229. The Pasha, 231.
Ardjen Lake, i. 385.
Armatoles of Olympus, ii. 46, 54.
Armenians, the, i. 7, 8.
Arta, valley of, ii. 183.
Ascent of the peak of Athos, i. 103. Of Ida, 16, 17. Of St. Elias on Olympus, ii. 14. Of Pelion, 123.
Atchi-keui, farm of, i. 44.
Athos, Mount, i. 3, 50; ii. 121. Plan of, i. 53. Monasteries, government of the, 69. Peak of, 55. Vegetation, 55. Scenery, 56. Climate, 57. Flowers on the peak, 106. Slavonic monasteries, 72, 123, 124. Exclusion of women, 63. Holy Synod, 63.
Avret Hissar, village of, i. 386.

B.

Babuna range, i. 158, 171, 352, 364, 377.
Bali-dagh, i. 34. Tumuli at, 34. Summit, 39. Mr. F. Calvert's excavations, 34.
Ballad of the Salamvria, ii. 71.

BALLADS.

Ballads of Modern Greece, ii. 224.
  M. Fauriel's collection, 226; Professor Ulrichs', 227.
Ballads of the Clefts of Olympus, ii. 51, 56.
Banja, village of, i. 378.
Barlaam, monastery of, ii. 158. Mode of ascent to, 157. Origin of the name, 158. Relics, 159. Ballad relating to the monastery, 162.
Barth, Dr., i. 342, 378, 381; ii. 23.
Bastrik, Mount, i. 340.
Beaujour's, M., account of Ambelakia, ii. 63.
"Beauty's Tower," The, ii. 67.
Bendscha River, i. 227.
Berat, city, i. 219. Population, 221. The Castle, 221.
—— to Corfu, i. 218.
Beratino River, i. 218.
Bertiscus Mountains, i. 285, 329, 330 *note.*
Beyramitch, town of, i. 9, 10, 21. Plain of, 38.
Bitolia, Christian name of Monastir, i. 166, 172 *note.*
Black Mountain, why so called, i. 237.
Black and White Drin, their confluence, i. 333, 334.
Black Wallachs, i. 351.
Bocche di Cattaro inlet, i. 235.
Boyana River, i. 279.
Brasidas, retreat of, from Lyncestis, App. C, ii. 361.
Brusa, city of, i. 1.
Bukova Monastery, i. 170, 174.
Bulgarian Church, i. 181.
Bulgarians, the, i. 176. Early history, 177. Bogoris, Christian monarch, 178. Samuel, King, extent of his conquests. 179. Their relation to the Greek Church, 182.
Bulgaro-Wallachian kingdom, ii. 179.
Bunarbashi, earthquake at, i. 6, 7. Springs at, 28. Homer's description of the springs, 30. Temperature of the springs, 32. River, 5, 36. Site of Troy on the hill above, App. A, ii. 355.

CHILANDARI.

Buthrotum, ancient, i. 233.
Butrinto, lake, i. 231.
Buyuk Magara, the Great Cavern, i. 21.
Byzantine pictures on Athos, i. 80, 100.

C.

Calcandele, town of, i. 352, 354.
Calliccolone, probable name of the tumulus of Æsytes, i. 40.
Calvert, Mr. F., i. 45.
Cambunian Mountains, ii. 20, 149.
Canal of Xerxes, i. 127. Isthmus through which it was cut, described by Herodotus, 127.
Cantacuzene, rebel, regent, and emperor, i. 138.
Caracalla, monastery of, derivation of name, i. 92 *note.*
Carya, plain of, ii. 40.
Caryes, or "the Hazels," i. 63, 129.
Castagneti, derivation of name of, i, 298.
Castelnuovo, town of, i. 235.
Casthanæa, ruins of, ii. 104.
Castri, birthplace of Scanderbeg, i. 298.
Catavothras, i. 159; ii. 185, 189.
Cattaro, bay of, i. 237. Town, 238.
Cavalla, town of, i. 51. Roman aqueduct at, 51.
Central basin of Olympus, ii. 17.
Cetinjé, i. 239. Plain of, 245, 253. Locanda, 253. M. Vaclik, Secretary to the Prince of Montenegro, 254. Neighbourhood, 266. The senate, 267. Crédit Mobilier, 267. The palace, 268.
Chaizi, port of, ii. 72.
Charon, ii. 327. Ancient and modern conceptions of, 329. Poems relating to, 243.
Chalcidice, trident of, i. 101.
Chiblak, village of, ruins near, i. 48.
Chigri, hill of, i. 21.
Chilandari, Bulgarian monastery of, i. 124. Greek MS. of St. John's Gospel at, 124. Intelligence of the leading monks, 124. First purely Slavonic monastery, 135.

'Childe Harold,' palace of Tepelen described in, i. 227. Description of Zitza, ii. 190.
Chiri River, i. 285.
Chiron, the centaur, cave of, ii. 125.
Christians of Scodra, their condition, i. 282.
"City," The, name applied to Constantinople, i. 97 *note*.
Classical superstitions existing among the Greeks, ii. 304. The Nereids, 320. The Genii, 317. The Lamia, 320. Gello, 321. The Three Fates, 321. Charon, 325. Christian use of pagan fables, 333.
Cleftic ballads, ii. 232, 258.
"Cleft's Arms," song of the, ii. 54.
Clefts of Olympus, ii. 50.
Cœnobite convents, i, 86.
Constamonitu monastery, i. 121. The hegumen, 121.
Constantinople, routes to, by sea and land, i. 2.
Corfu, routes to, from Salonica, i. 149.
Croia mountains, i. 222, 317.
Customs and beliefs of the Wallachians, ii. 172.
Cutlumusi, monastery of, i. 73.
Cyril and Methodius, i. 191.
Czerna River, i. 376.
Czerna Gratzko, i. 376. Probable site of Stobi, i. 377.

D.

Dalmatia and Bosnia, i. 236.
Danae, fable of, ii. 288.
Dante, description in 'Il Paradiso' applied to monastic contentment, ii. 35.
Dardanelles, town of, i. 2.
Delvino, town of, i. 231.
Demetrias, ruins of, ii. 130.
Demirkapu, or Iron Gate of the Vardar, i. 379.
Dereli, village of, ii. 59.
Derven Aga, guardian of the mountain passes, ii. 7, 144.
Devol River, i. 208. Valley of the, 218.

Dionysius, S., monastery of, on Olympus, ii. 13.
———, confounded with Dionysus, ii. 13. Tradition relating to, 14.
Dium, ruins of, ii. 7.
Djudan, marsh and lake, i. 45.
Docheiareiu, or "the Steward's Monastery," illuminated manuscript at, i. 119.
Dodona, site of, App. G, ii. 378. Hesiod's description of the neighbourhood, 378.
Doiran, ancient Tauriana, i. 384.
Don Nicola Bianchi, priest of St. George, i. 294.
Dramisius, Greek theatre at, ii. 201. Ruins of an Hellenic city, 203.
Drin River, i. 289, 333.
Dryno River, i. 228.
Ducadjini, laws of the, i. 308.

E.

Eastern Church, distinction between statues and pictures in, i. 193. Dean Milman on, 194.
——— Christendom, tolerance of, i. 105.
——— monastic life, i. 68.
Edessa, ancient, i. 156.
Egnatian Way, the, i. 149. Appendix D, ii. 363. Strabo's description, 363. The Itineraries, 363. Tables of distances, 364. Trajectus Genusi, 367. Pons Servilii, 367. Lychnidus, 368. Road between Edessa and Heraclea, 369.
Elassona, mountain of, ii. 36.
Elbassan, city of, i. 201. Identification with Scampæ, 202 *note*. Religion of the inhabitants, 202. The Ghegs, 205. Route from, 206.
"Elympos," i. 152 ; ii. 19.
Enipeus River (the ancient), ii. 29 and *note*.
Enaeh, ancient Neandria, i. 9.
Episkopi, hill of, ii. 127.

EUCHARISTIC.

Eucharistic cakes, i. 92.
Eugenius Bulgaris of Corfu, school founded by, i. 62.
Eurymenæ, site of, ii. 78.
Evjilar, village of, i. 11. 21.

F.

Fallmerayer, German historian, i. 89, 182. His description of the Wallachians at the present time, ii. 180.
Fauriel, M., Chants Populaires, ii. 226. Account of a myriologue, 242.
Fendroudi, village of, i. 218.
Festival of the Transfiguration on the summit of Athos, i. 99, 104.
"First Man," office of, i. 65, 134.
Future of the Holy Mountain, i. 132.

G.

Galliko River, the ancient Echidorus, i. 386.
Gargarus, mount, i. 16.
Garlik, hill, i. 44.
Gergithus, suggested site of, i. 43.
Ghegs and Tosks, i. 213.
"Giants' Mountains," the, ii. 135.
Glyky, Inferno of the Greeks, ii. 219.
Gonnus, site of, ii. 61.
Goritza, Christian suburb of Berat, i. 218. Hill of, ii. 129.
Gossip on Athos, i. 109.
Gradet Khan, the, i. 381.
Gradiska, village of, i. 382.
Gratschan, village of, i. 375.
Greek nursery rhyme, ii. 260.
Grisebach, Dr., description of the Mirdite country, i. 290, 329.
Gurnitzovo, Christian village, i. 162.

H.

Hagion Oros, or Monte Santo, i. 54.

IVERON.

Hanai Tepe, tumulus, i. 45.
Heequard, M., on the history of the Concealed Christians, i. 346.
Hellespont, the, i. 3, 38, 40.
Hereditary Pashas, i. 356. Dr. Grisebach on, 356.
Hermit, a, i. 120.
Hesiod, battle of the Gods and Titans described by, ii. 45.
Hesychasts, the, i. 139.
Hissarlik, ruins, i. 48.
History of the Monks on Athos, i. 133.
——— of Salonica, i. 147.
Holy Mountain, the general features of, i. 52. Life on, 60. Treatment of, by the Latins, 136. Later history of, 141.
Holy Synod of Athos, i. 63.
Homeric topography, i. 23-28.
Homoloium, probable site of, ii. 72.
"Honey Mount," i. 37.
Hotti and Clementi, Albanian tribes of, i. 277.

I.

Ida, Mount, i. 1. Pitch of, 14. Bivouac on, 15. Sacred character of, 16. View from its summit, 18. Flora of, 18.
Idiorrhythmic Convent, i. 69, 87.
Ilaian Plain, i. 49.
Ilium Novum, ruins of, i. 48.
Indian tales, ii. 269, 277.
Innocent III., i. 136.
Iolcos, site of, ii. 128.
Ionians, the, i. 68.
Ipek, tableland of, i. 359.
Ismael Pasha, Governor of Scodra, i. 284.
Isthmian pine, i. 124.
Isthmus of Pallene, canal through, i. 99.
Ivan the Black, i. 246.
Iveron, convent of, i. 76. Architecture, 77. Central Church, 78. Library, 82.

J.

Jabokika, village of, i. 225.
Justinian, aqueduct attributed to, i. 369.
———, birthplace of, i. 372. App. E, ii. 370. Ochrida, 370. Justiniana Prima, 371. Gibbon's account of, 371. Uskiub, the ancient Scupi, 372.

K.

Kako Suli, village of, ii. 210.
Kalabaka, ancient Æginium, ii. 149. Defeat of the Greek insurgents at, 165.
Kanalia, monastery of, ii. 33.
Kara-dagh, the, or Black Mountain, i. 352, 368.
Karitza, village of, ii. 78.
Karla, lake of, ancient Bœbe, ii. 109.
Katrin, village of, ii. 4, 141.
Keramidi, ruins near, ii. 103, 105.
Khan of Jura, i. 200.
Khortiatzi, Mount, i. 143; ii. 21.
Kiuprili, i. 372.
Kimar River, i. 38.
"Kirke Gheuz," or "the Forty Eyes," i. 22.
Kinurio, or New-place, i. 232.
Kobelitza, peak of, i. 353, 360.
Koraphia, Mount, i. 340.
Kukus, village of, i. 200.

L.

Larissa, ii. 44, 143. The ex-Dervenaga's residence at, 144.
"Las Incantadas," monument at Salonica, i. 148.
Laura of Scetis, Mr. Kingsley's description of, applicable to Retreat on Mount Athos, i. 98.
Lavra, or Laura, monastery of, i. 93. Relics at, 95.
Lavamani, or Concealed Christians, i. 345.

VOL. II.

Lectum, Cape, i. 18.
Leftocarya, village of, ii. 31.
Legends of Mount Athos, i. 102.
Lekhonia, town of, ii. 134.
Lemnos, island of, i. 35, 93.
Lepenatz River, i. 361.
Letochoro, village of, ii. 7, 29.
"Lidja, the," &c. "the Refuge," cavern, i. 15.
Livadaki, Mount, ii. 41.
Livari, village of, i. 232.
Lövchen, peak of, i. 237. Burial-place of Peter II., the last Vladika, 244.
Lower Olympus, ii. 29. Geological character of, 32. Western side of, 37.
Lycanthropy among the Greeks, ii. 83.
Lykostomo, or "the Wolf's Mouth," site of Gonnus, ii. 60.

M.

'Maiden in Hades,' the ballad of, ii. 327.
McLennan on 'Primitive Marriage,' i. 319.
Magnesia, district of, ii. 114.
Mahometan cemeteries, i. 10, 201.
——— protection of the monks on Mount Athos, i. 140.
Makrinitza, village of, ii. 127.
Malacassi, village of, ii. 164. Destruction of, in 1854, 164.
Malaria fever in Turkey, i. 223.
Maritza River, i. 336, 340.
Max Müller, theory of Mythology, ii. 288 and *note*.
Melchizedeck, of the Lavra, i. 94.
Melibœa, ancient, ii. 99.
Mendere River, i. 8, 11, 36, 37, 44. App. A, ii. 340.
Meteora, rocks of, ii. 149. Great Monastery of, 150. Convent of St. Stephen, 152. Barlaam, 156.
Metrophanes Critopulus, i. 141.
Metzovo, town of, ii. 169.
Mézières, M., on the Pelion fortresses, ii. 100.
Metzikeli, Mount, ii. 186.

2 C

MILIES.

Milies, village of, ii. 136. Public library at, 138.
Miraculous picture at Iveron, i. 83.
Mirdita, i. 292. Oak-forests of, 293. Churches and priests, 294, 296. Metals, 296. Rivers, 299. Palace of Orosch, 300. Prince of, 302. History of his family, 303. Ravages of the vendetta, 305, 310. Political constitution, 307. Administration of justice, 309. Monte Santo, 314. Topography, 316 and *note*.
Mirdite shepherds' encampment, i. 331.
Mirdites, the, i. 292. Dress of, 293. Religious opinions of, 297. Fraternal friendships, 309. Derivation of the name, 313. Capture of wives, 318. Wives by birth Mahometans, 325.
Modern Greeks, their origin, ii. 304. Pagan superstitions, 307. The Nereids, 309. Genii, 317. Grotto at Chios, 319. State of the dead, 323. Various customs, 331.
Moglena, district of, i. 384.
Moglenitiko River, i. 155.
Molossia, district of, ii. 203.
Molossian dogs, ii. 203, 204 *note*.
Monasteries in Wallachia and Moldavia, i. 70.
Monastic dispute at Athos, i. 66.
—— life, the Cœnobite and Idiorrhythmic rules, i. 69.
Monastir, city of, i. 166. Military importance of, 167. Population, 167. Parade-ground, 167. Greek inscription found at, App. B, ii. 358.
——, plain of, i. 165, 171.
—— to Ochrida, i. 183.
Monks of Athos, i. 71, 84, 131.
Monks' views of other Christian churches, i. 125.
Montenegrin Church, i. 262. Dress, 242.
Montenegrins, the, i. 272.
Montenegro, i. 234. Boundary of territory, 238. Approach to, 241. History of, 245. Struggles with

OLYMPUS.

the Turks, 248. Recognized by Peter the Great, 249. Stephen the Little, 250. Two last Vladikas, 251. The capital of, 253. Political constitution, 255. The old régime, 256. Population and revenue, 257. Trade. 258. Monastery, 259. The Archimandrite, 261. National songs, 265. Senate of, 267.
Monte Santo in Albania, i. 313. Scala Santa, 314.
Motives for leading monastic life at Athos, i. 88.
Mysian Olympus, flora of, i. 18.
Mysticism among the Orientals, i. 90, 105, 139.

N.

Neapolis, i. 52.
Negotin or Tikvesh, i. 377.
Nereids, the, ii. 309. Professor Ross on, 310. Ballad on, 313. Origin of the superstition, 315.
Nezero, lake of, ii. 40. Village of, 41. Origin of the name, 42.
Nicene Creed, a common standard of faith, i. 125.
Nicolas, Prince of Montenegro, i. 268.
Nidjé, Mount, i. 158, 164, 364.
Niégush, village of, i. 242.
Norse and Greek tales, ii. 273.
Nymphæum, the ancient, i. 108.

O.

Ochrida, Lake of, i. 185 ; boats on, 196. City of, i, 185. Inhabitants, 186. Metropolitan Church of, 187. Byzantine crucifix, 187.
Ochrida to Elbassan, i. 195.
Olaus Magnus, i. 322.
Olympian insurrection of 1854, ii. 46.
Olympus, i. 118, 152, 156, 385 ; ii, 1, 5. Ascent of, ii, 8, 14. Sawmills of, 11. Map of, 17. Derivation of the name, 19 and *note*. Highest summits, 25. Descent

OLYTZIKA.

of, 27. Mixture of races on, 42.
National ballad, 51. March of a
Roman consul across, App. F,
374. Livy's narrative, 374. Topography, 375.
Olytzika, Mount, ii. 203, 206.
Oriental treatment of women, ii. 119.
Orosch, i. 299, 302. Palace of, 300.
Parish church at, 318.
—— to Prisrend, i. 327.
Ossa, ii. 40, 70, 72, 78, 98, 142.
Ostrovo, village of, i. 158. Mosque, 158. Lake, 159. Subterranean channels, 159. Legends, 160. Pliny and Catullus, 161. Waddington, 161.
Othrys, heights of, ii. 45, 109, 124.
Ottoman government, i. 163, 175, 207, 282, 394; ii. 2, 102, 132.

P.

Palace of Zeus on Olympus, ii. 22.
Palæologi, the, i. 137.
Palus Acherusia, ii. 219.
Panselenus, Byzantine artist, i. 80.
Pantocratoros, or The Almighty, monastery of, i. 73.
Parga, ii. 220. Sir A. Alison's account of its cession, 221. Greek ballad of, 223.
Parnassus, ii. 124, 136.
Pelagonian Plain, ancient, i. 172.
Pelion, ii. 98. Vegetation on, 121. Summit of, 123. View from the summit, 124. Ridge of, 106, 109.
Pella, site of, i. 153. Birthplace of Alexander the Great, 154.
Peneius, ii. 60. Upper valley of, 163.
Perasto, town of, i. 237, 240.
Pergamus, summit of, i. 41.
Peristeri, Mount, i. 164, 168, 364.
Perlepe, town of, i. 171; ii, 90.
Philotheu Monastery, i. 84, 91.
Pieria, city of, ii. 6. District of, 8, 20, 32.
Pierus, Mons, ii. 6.
Pigs in Turkey, i. 162.

RIVERS.

Pindus, ii. 163. Range of, 37, 146.
Continuation of Scardus, i. 184.
Plain of Troy, map of, i. 23.
Plamenatz, Pope Elia, warrior of the Black Mountain, i. 254.
Platamona, castle of, ii. 31; site of ancient Heracleium, ii. 31, 141.
"Politarchs," i, 145 and *note*.
Polydendron, ii. 101. Woods of, 102.
Popular Tales, Modern Greek, ii. 261. Von Hahn's collection, 262. Beast Fables, 265. 'The Wolf, the She Fox, and the Pot of Honey,' 267. In different countries, 269. 'Cinderella,' 270. 'The Snake Child,' 271. Views as to their origin, 276. Indo-European origin, 281. The Drakos, 293, 301. 'Lazarus and the Dragons,' 295. Drakos mythological, 299. 'Jack and the Drakos,' 303.
Portaria, village of, ii. 127.
Prenk Bib Doda, the Mirdite Prince, i. 303. History of the family, 303.
Presba, plain and lake of, i. 184.
Prisrend, i, 290, 336. Dress of the inhabitants, 337. Nazif Pasha, 338. Castle of, 339. Churches, 341. Archbishop of, 343. Population, 344.
Prisrend to Uskiub, i. 350.
Provlaka, canal of Xerxes, i. 127.
Punto d'Ostro, i. 235.

R.

Railway route across Turkey, i. 387. Von Hahn's opinion, 387.
Read, Mr., British Consul at Scodra, i. 281.
Retreat of St. Demetrius, i. 99. Greek artist at, 100.
—— of "the Forerunner," i. 97.
Retrospective view of the monasteries on Athos, 130.
Rhœtean promontory, i. 49.
Rieka, town, i. 270.
Rivers of Greece and Asia Minor, i. 8.

ROMAIC.

Romaic ballads, ii. 224. Subjects of, and leading characteristics, ii. 232. Homeric features, 235. Idyllic pieces, 237. 'Bridge of Arta,' 239. 'The Abduction,' 240. Dirges or myriologues, 241. Love poems, 245. 'Demos,' 'The Garden,' 247. Distichs, 248. "Political" verse, 251. 'The Young Sailor,' 256.
Romaic and Neo-Hellenic languages, ii. 116.
Romaika, dance, ii. 118.
Roman colonies in Dacia, ii. 175.
Route Impériale from Salonica to Monastir, i. 150, 157, 165.
Rumia, peak of, i. 244.
Rumuni, name appropriated by the Wallachs, i. 224 ; ii. 171.
Russico, or the Russian monastery on Mount Athos, i. 129.

S.

St. Athanasius, the founder of the Lavra monastery, i. 94, 97, 102, 134.
St. Clement, legend of, i. 188. Statue of, 187.
St. Demetrius, monastery of, ii. 72. Date of foundation, 74. Plan of the church, 75. Architecture, 76.
St. Dionysius, monastery of, on Athos, i. 111. Relic described by Mr. Curzon, 112. Monastery of, on Olympus, ii. 9. Legend of, 12.
St. Elias, peaks named from, i. 16 and *note* ; ii. 20, 37, 41.
St. Gregory's monastery, i. 113.
St. Naum, monastery of, i. 174.
St. Nicolas, island of, ii. 139.
St. Panteleemon, monastery of, scene of Ali Pasha's assassination, ii. 192.
St. Paul, monastery of, i. 109. Meal at, 110.
Salonica, i. 143. Triumphal arches, 144. Ecclesiastical antiquities, 145. Inhabitants, 146. History

STRUGA.

of, 147. Bad roads, 150. Bay of, 388. Cholera at, ii. 1.
Salonica to Monastir, i. 143.
Samothrace, i. 50.
Samuel the Caloyer, ii. 213.
San Giorgio, i. 294, 298.
Sand-bath, the, i. 75.
Sane, ancient town of, i. 128.
Sarigöl, Lake, i. 162.
Scala of Cattaro, i. 238.
Scamander, sources of the, i. 20.
Scamandria, supposed site of, i. 43.
Scamnia, village of, ii. 37, 40.
Scanderbeg, i. 215. Ballad on his death, i. 217. Birthplace of, 298.
Scardus range, i. 336, 350. Passes of, 184, 351. Flora of, 352. Country east of, 363. Country west of, 364.
Scodra, lake of, i. 244, 277, 298. Mode of catching the scoranzi, 278. City and bazaars, 280. Population, 280. Castle, 284. Sieges, 286.
Sedjin, village of, i. 328.
Sepias, cape, ii. 111.
Servian newspapers, i. 264.
Siegel, Professor, Bœotian story relating to S. Dionysius, discovered by, ii. 14.
Simois River, i. 40 ; ii. 345.
Simopetra, monastery of, i. 113. Inmates of the, 115.
Site of ancient Troy, i. 39. See Appendix A, ii. 337.
Skete of St. Anne, i. 108. Wood carvers at, 108.
Skiti, ruins at, ii. 98.
Skumbi River, i. 200, 207.
Slavonic names in Albania, i. 225.
"Slayer of the Bulgarians," i. 180.
"Sleeping Beauty," the, ii. 283. Mythological significance of, 285.
Sopoti, Mount, i. 231.
Sparmos, monastery of, ii. 37.
'Spectre,' the, ballad on the Vrykolakas, ii. 95.
Sphigmenu Monastery, i. 128.
Stavroniceta, monastery of, i. 75, 141.
Stephen Dushan, i. 137.
Struga, town of, i. 195. Bridge over

STYX.

the Drin at, 197. Roman milestone at, 198. Bulgarian school at, 199.
Styx, in Arcadia, i. 26.
Suli, district of, ii. 199. Mountains of, 207. Hero of, 213.
Suliotes, the, ruined homes of, ii, 209. Their history, 210. Ballad of Despo, 216.

T.

Temenidæ, the legend of, i. 173.
Tempé, Pass of, ii. 66. Ancient descriptions, 69.
Tempé and Ossa, ii. 59.
Tettovo, or district of Calcandele, i. 359.
Tepelen, birthplace and residence of Ali Pasha, i. 227. Palace described in 'Childe Harold,' ib.
Thasos, Island of, i. 50.
Therike, ii. 206.
Therma, city of, i. 147.
Thessalian Olympus, i. 26.
Thessalonica, i. 147. Sieges of, 148.
Thessaly, ii. 66, 80. Mountains and plains, 44, 108, 142, 152. Greek insurrection in, 166.
—— and Meteora, ii. 140.
"Three peaks of Heaven," ii. 24 note.
Tmolus, Mount, i. 17.
Tomb of Hector, view from, i. 35.
Tomohr Mount, i. 203, 219.
Tournefort's narrative of the Vrykolaka superstition, ii. 92.
Trees, places named from, ii. 107.
Tricala, ii. 147. Ancient name of, 148.
Trikeri, ancient Aphetæ, ii. 124.
Trojan Plain, i. 4, 36. Its inundations, 37.
Troy, topography of, i. 23. Appendix A, ii. 337. Bali-dagh, Atchi-keui, Hissarlik, i. 39; ii. 353. The Mendere, 341. Bunarbashi River, 343. The Dumbrek, 345. The Kimar, 347. Kumkaleh, 348. Position of the

VOLO.

Greek Camp, 349. Pursuit of Hector by Achilles, 351.
Tumuli on the Bali-dagh, i. 34.
—— of Achilles and Patroclus, i, 5.
Turcoman encampment, i, 19.
Turkey, migration of labourers, i, 389. Commercial Treaty with England, 391. Eastern question, 393. Greek and Slavonic races, 395. Future prospects, 397.
Turkish menzil, or post system, i. 195, ii, 141. Outrages, i. 175.
Turks and Montenegrins, i. 283.

U.

Ujek-tepe, tumulus, i. 40.
Uskiub, history of the city, i. 361. Situation, 366. Population, 366. Clock tower, 368. Justinian's Aqueduct, 369.
—— to Salonica, distances in hours, i. 386 note.
Usumi Valley and River, i. 218.

V.

Valonia oak, i. 9.
Vampires, malignant, ii. 85. Innoxious, 95.
Vardar, or Axius, river, i. 151, 360. Khan, or inn, on the banks of, i. 152. Valley, 371. Lower part of, 374. Unexplored route, 375. Iron gate of, 379. Lower course of the river, 383. Gate at Salonica, 144.
Vatopedi Monastery, i. 52, 59. Architectural antiquity of the central church, 62. Relics, 62. Inmates, 62.
Veluki, peak of, ii. 136.
Viosa River, i. 226.
Vistritza River, i. 231.
Vodena, city, i. 155. View from, 156.
Volo, town of, ii. 131. Roman aqueduct, 131. The Lazzaretto, 131.
—— Gulf of, ii. 124.

VON.

Von Hahn, excavations by, on the Bali-dagh, i. 42.
Vrykolaka, the, or Eastern Vampire, ii. 80. Derivation of the name, 82. Anecdotes of the superstition, 88. Modes of exorcism, 91.

W.

Wallachs, the, i. 167, 224, 351; ii. 16, 43, 169, 237. Their history, 174. Present condition, 180.
—— of the Balkan, ii. 178.
War of Independence, ii. 58.
"Wolf's Mouth," or Pass of Tempé, the, ii. 66.

X.

Xenophu, monastery of, i. 118. Iconostase in the Church, 119.
Xeropotamu, or "the Torrent," i. 116, 133. Fragment of the true cross, 116. Cup of the Empress Pulcheria, 117.
Xerxes, Canal of, i. 55.

ZYGOS.

Y.

Yanina, ii. 185. Population, 186. Probable site of Dodona, 186, and Appendix G, ii. 378.
Yanina and Zitza, ii. 183.
Yenidje, town of, i. 154.
Yenishehr, village of, the ancient Sigeum, i. 4.
Yuruks and Turcomans, origin and distinction of, i. 13.

Z.

Zadrima, plain of, i. 290.
Zagora, town of, ii. 111. Its commerce, ii. 113. Church and Festival of the Transfiguration at, 115. View of Athos from, 121.
Zara, Montenegrin political refugees sent to, i. 240.
Zaribaschina, peaks of, i. 353.
Zitza, monastery of, ii. 186. Byron's visit to, 189; and description, 190.
Zographu, or "The Painter's Monastery," i. 123.
Zygos, pass of, ii. 167.

THE END.

PRINTED BY W. CLOWES AND SONS, DUKE STREET, STAMFORD STREET,
AND CHARING CROSS.

BOOKS OF TRAVEL.

ART OF TRAVEL; or, Hints on the Shifts and Contrivances available in Wild Countries. By FRANCIS GALTON, F.R.G.S. Third Edition. Woodcuts. Post 8vo. 7s. 6d.

A MANUAL OF SCIENTIFIC INQUIRY, prepared for the Use of Officers and Travellers in general. Third Edition. Post 8vo. 9s. (By Order of the Lords of the Admiralty.)

REMINISCENCES OF ATHENS AND THE MOREA: Extracts from a Journal during Travels through Greece in 1839. By the late Earl of CARNARVON. Edited by the present Earl. Map. Crown 8vo. 7s. 6d.

VOYAGE OF A NATURALIST ROUND THE WORLD. By CHARLES DARWIN, F.R.S. Illustrations. Post 8vo. 9s.

TRAVELS IN THE ISLANDS OF THE EAST INDIAN ARCHIPELAGO; a Popular Description of their Natural History and Geography, with some Account of Dangers and Adventures among many Tribes. By ALBERT S. BICKMORE, F.R.G.S. With Maps and Illustrations. 8vo. 21s.

THE NILE AND ITS BANKS; showing their Attractions to the Naturalist, Archaeologist, and General Traveller. By Rev. A. C. SMITH. With Woodcuts. 2 vols. Post 8vo. 18s.

VISITS TO THE MONASTERIES OF THE LEVANT. By the Hon. ROBERT CURZON. With 18 Illustrations. Post 8vo. 7s. 6d.

MISSIONARY TRAVELS AND RESEARCHES IN SOUTH AFRICA. By DAVID LIVINGSTONE, M.D. With Map and 30 Illustrations. Post 8vo. 6s.

BUBBLES FROM THE BRUNNEN OF NASSAU. By the Right Hon. Sir FRANCIS HEAD, Bart. With 13 Illustrations. Post 8vo. 7s. 6d.

LETTERS FROM HIGH LATITUDES; an Account of a Yacht Voyage to Iceland, Jan Mayen, and Spitzbergen, &c. By Lord DUFFERIN. With 24 Illustrations. Post 8vo. 7s. 6d.

NINEVEH AND ITS REMAINS. A Narrative of an Expedition to Assyria in 1845, 1846, and 1847. By the Right Hon. AUSTEN H. LAYARD, M.P. With 100 Illustrations. Post 8vo. 7s. 6d.

NINEVEH AND BABYLON. A Narrative of a Second Expedition to Assyria in 1849, 1850, and 1851. By the Right Hon. AUSTEN H. LAYARD, M.P., D.C.L. With 150 Illustrations. Post 8vo. 7s. 6d.

[*Continued.*

BOOKS OF TRAVEL.—*continued.*

THREE YEARS' RESIDENCE IN ABYSSINIA, with Travels and Adventures in that Country. By Mansfield Parkyns. With Map and 30 Illustrations. Post 8vo. 7s. 6d.

THE RIVER AMAZONS. A Record of Adventures and Travels, with Sketches of Brazilian and Indian Life. By H. W. Bates, F.R.G.S. With Map and 40 Illustrations. Post 8vo. 12s.

BRITISH COLUMBIA AND VANCOUVER ISLAND. Their Forests, Rivers, Coasts, and Gold Fields, and Resources for Colonisation. By Commander Mayne, R.N. Illustrations. 8vo. 16s.

LAST WINTER IN AMERICA. Being Table Talk collected during a Tour through the late Southern Confederation, the Far West, the Rocky Mountains, &c., &c. By the Rev. F. Barham Zincke. Post 8vo. 10s. 6d.

TRAVELS IN PERU AND INDIA, for the purpose of collecting Cinchona Plants, and introducing Bark into India. By C. R. Markham. Illustrations. 8vo. 16s.

A RESIDENCE IN BULGARIA; or, Notes on the Resources and Administration of Turkey—the Condition and Character, Manners, Customs, and Language of the Christian and Mussulman Populations, &c. By S. G. B. St. Clair and C. A. Brophy. 8vo. 12s.

THE GREAT SAHARA. Wanderings South of the Atlas Mountains. By Rev. H. B. Tristram, M.A. Illustrations. Post 8vo. 15s.

TRAVEL AND ADVENTURE IN THE TERRITORY OF ALASKA (RUSSIAN AMERICA), and in other parts of the NORTH PACIFIC. By Frederick Whymper. With Map and 30 Illustrations. 8vo. 16s.

TWO YEARS' RESIDENCE IN DENMARK, JUTLAND, and COPENHAGEN. By Horace Marryat. Illustrations. 2 vols. 8vo. 24s.

A YEAR IN SWEDEN, including a Visit to the Isle of Götland. By Horace Marryat. Illustrations. 2 vols. Post 8vo. 28s.

SINAI AND PALESTINE, in Connection with their History. By Canon Stanley. Sixth Edition. Plans. 8vo. 16s.

FIVE YEARS IN DAMASCUS. With Travels to Palmyra, Lebanon, and other Scripture Sites. By Rev. J. L. Porter. Illustrations. Post 8vo.

TRAVELS IN EGYPT, NUBIA, SYRIA, AND THE HOLY LAND. By Irby and Mangles. Post 8vo. 2s.

PORTUGAL, GALICIA, AND THE BASQUE PROVINCES. By Lord Carnarvon. Post 8vo. 3s. 6d.

JOHN MURRAY, ALBEMARLE STREET.

www.ingramcontent.com/pod-product-compliance
Lightning Source LLC
Chambersburg PA
CBHW020740020526
44115CB00030B/721